Adventures in Mostly Calculus Mathematics

by

Richard J. Palmaccio

Copyright © 2012

All rights reserved

v. 1.2

Dedicated to

All my students past and present.

Note to the Reader

This book is readable by those who have studied first year calculus. It consists of problem adventures, many of which originate with the author. Most require some understanding of elementary calculus and provide some intriguing applications in many widely diverse areas. An examination of the Table of Contents will make that obvious. The author has taught courses in mathematics including Advanced Placement calculus AB and BC for the past 46 years and as of this writing is still active. This work is intended to share some time tested ideas with others and to enjoy the fun of discovering some intriguing applications.

Table of Contents

Note to the Reader	4
Highway Design Optimizing Automotive Fuel Economy	7
The Rotating Mercury Reflector	10
Crossing a Flowing River in Least Time	18
The Radius of a Black Hole	23
The Two Cannon Projectile Scenario	27
Modeling Average Temperatures with Complex Numbers	31
A Curious Recursive Sequence and a More Curious Nth Term Formula	38
Aircraft Surveillance of Speeders – Realistic Case	50
An Artillery Challenge	55
Proving the Parabola and Ellipse Focusing Properties Without Calculus!	59
A Curious Fountain Problem	63
The Volume of a Hypersphere	69
Newton's Law of Heat Transfer and Measuring Very Hot Objects	74
Cooling Tea and Newton's Law of Heat Transfer	77
Kepler's Laws: The Radius Vector Sweeps Out Equal Areas in Equal Time Increments and All Orbits are Conic Sections	87
The Derivation of a Pursuit Curve	95
Optimizing a Storage Area	101
A related Rates Approach in Determining Galactic Rotation	107
Optimizing the Gathering of Solar Energy on the Rooftop	119
A Fluid Flow Application of Linear 2nd Order Differential Equations	124
Time Dilation Explained	128
The Weight Watcher Function	131

Highway Design Optimizing Automotive Fuel Economy

Introduction

The author came across a problem in *Essential Calculus* by James Stewart which is a challenge problem, number 18, in Chapter 4. It deals with Poiseuille's Law which basically states that the rate of blood flow through a blood vessel is proportional to the fourth power of its radius. Other factors such as pressure and viscosity enter into the exact law. Stewart's problem asks the solver to determine the best branching angle whereby a smaller vessel branches off of a larger vessel so that there is a minimum of resistance to the flow. This caused the author to think about traffic flow on highways. A rough analogy to viscosity and flow resistance is the average number of gallons per mile of fuel consumption. The problem in this article is the result.

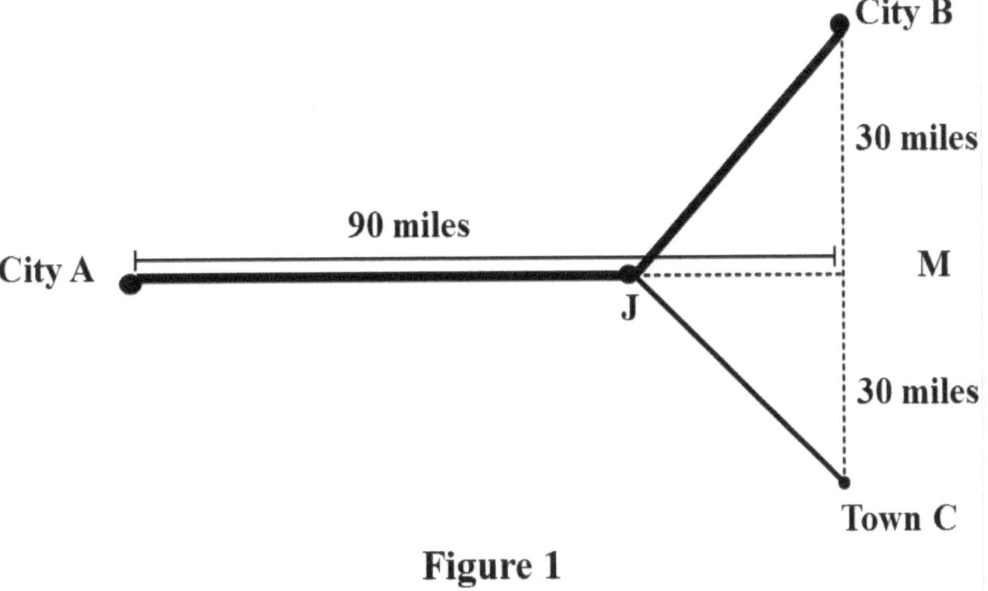

Figure 1

The Problem

City A is located 90 miles due west of the midpoint, M, of a 60 miles long north-south line joining City B to Town C as indicated in Figure 1. A highway is to be built together with a branching smaller road at point J. Studies indicate that on the average day 60% of the traffic will flow between City A and City B while the remaining 40% will travel between City A and Town C. Due to the nature of the terrain and fewer lanes of traffic, cars on the J to C portion of the roadway will average 21 miles per gallon while the faster travel allowed on the remainder of the road system will allow cars to obtain 30 miles per gallon. Where should the junction at J be placed so as to minimize the fuel use of the traffic pattern described?

Let θ be the measure of angle BJM = measure of angle CJM. Let N = the number of cars traveling on the A to J portion of the highway. Figure 2 contains some necessary information. With θ as defined, JM = 30 cot θ so that AJ = 90 − 30 cot θ. JC and JB are each 30 csc θ in length.

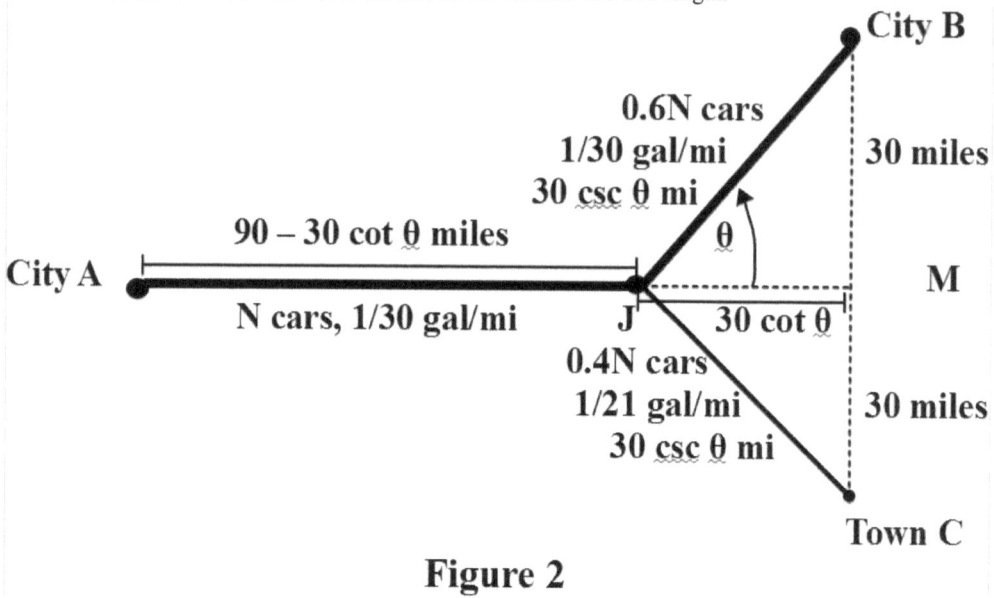

Figure 2

Fuel usage is inverted so that 30 mpg becomes 1/30 gallons per mile. This enables multiplication of gal/mi by miles to obtain the number of gallons consumed by each car. Then multiplying by the distance driven calculates the number of gallons of fuel consumed by the traffic pattern. Thus the total fuel use is given by the function below.

$$F(\theta) = \frac{N}{30}(90 - 30\cot\theta) + \frac{.6N}{30}(30\csc\theta) + \frac{.4N}{30}(30\csc\theta)$$

This simplifies to

$$F(\theta) = N\left(3 - \cot\theta + .6\csc\theta + \frac{4}{7}\csc\theta\right)$$

To obtain the domain of this function we think of point J moving from point A to point M. At point M, θ = 90° while at point A we have θ = tan⁻¹(1/3) which is approximately 18.4°.

Differentiating yields

$$F'(\theta) = N\left(\csc^2\theta - \frac{82}{70}\csc\theta\cot\theta\right)$$

8

In addition to the critical values of 18.4° and 90° we set the derivative equal to zero to obtain any additional critical values.

$$F'(\theta)=0$$
$$\csc\theta = \frac{82}{70}\cot\theta$$
$$\cos\theta = \frac{70}{82} = \frac{35}{41}$$
$$\theta = \cos^{-1}(70/82) \approx 31.39°$$

This result suggests that $90 - 30\cot\theta = 71.7$ miles from A might be the optimal place to locate the junction so as to minimize fuel use. This needs to be verified. We use the second derivative test.

$$\frac{1}{N}*F''(\theta) = -2\csc^2\theta + \left(\frac{41}{70}\right)*\left(\csc\theta\cot^2\theta + \csc^3\theta\right)$$

$$\frac{1}{N}F''(\theta) = \frac{-29}{35}\csc^3\theta + \frac{41}{35}\csc\theta\cot^2\theta$$

It can be easily determined that $F''(31.39)$ is positive, proving that fuel use is at its minimum for the 71.7 mile placement of the junction away from City A.

The Rotating Mercury Reflector

Introduction

You know that the parabola is the set of all points equidistant from a given fixed point (the focus) and a given fixed line (directrix). If you rotate a parabola about its axis of symmetry a surface called a *paraboloid* is formed. This surface is the set of all points in space equidistant from a fixed point and a fixed plane (the plane of all positions of the directrix as it rotates about its intersection with the axis of the original parabola). Such paraboloids are employed as satellite dishes, flashlight or auto headlight reflectors, and the like.

Here is one of those seemingly "made in heaven" coincidences that makes mathematics so fascinating to math fans. If you rotate a circular container of liquid such as a basin full of water, or even a cup of coffee, the surface of the liquid assumes the shape of a paraboloid!!! The paraboloid is only degraded very near the edge of the container as surface tension disturbs an elegant balance between the centrifugal force of rotation and the force of gravity. The subject of this article is the proof of this fantastic property of rotating liquids.

The title refers to one of the most exquisite applications of this property. Mercury, that liquid metal, is more perfectly reflective than an ordinary mirror (provided oxidation with the air is avoided). A reflective telescope requires a large parabolic mirror to gather light and focus it to a point where a flat mirror sends the light through a few lenses forming the eyepiece of the telescope. Large glass mirrors take months to make and must be very nearly perfect. A rotating mercury reflector telescope contains a large shallow basin rotated at a precise angular velocity which controls the location of the focus. Figure 1 shows such a mercury basin and the rays show representative paths of light rays converging at the focus.

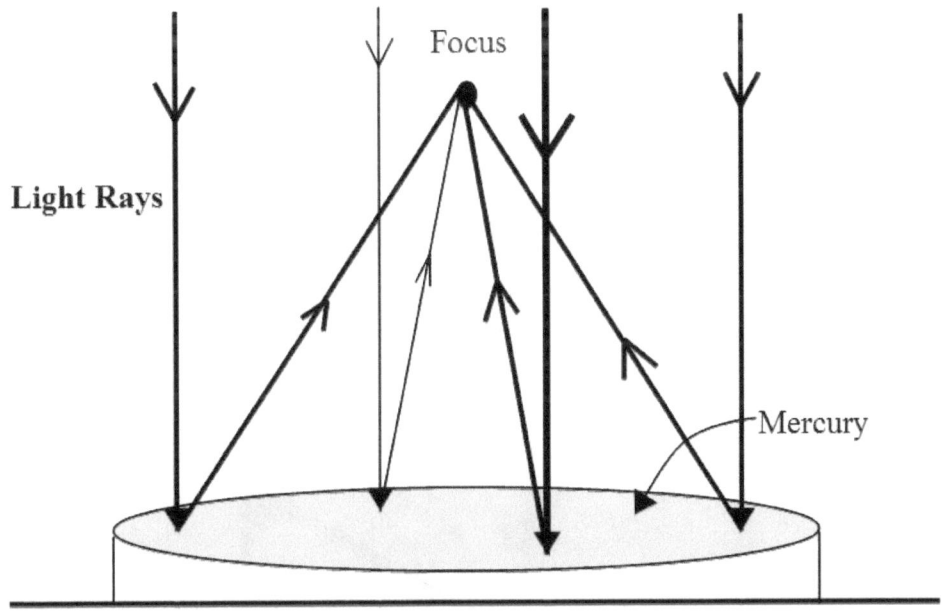

Figure 1

You can experiment yourself by taking a large basin half-filled with water in a darkened room with a smooth ceiling. Stir the water until you get a good depression at the center. Then take a point source of light, such as a lighted match (be careful) and hold it over the center of the basin. You should then move the light up and/or downward until the light is at the focus. You should see a bright reflection on the ceiling. If you hold the light stationary (watch your fingers), as the rotation of the water gradually slows, the focus moves upward and the image widens unless you raise the light.

Some Vector Preliminaries

You need just a tiny amount of background in vectors in order to enhance the understanding of why rotating a container of liquid causes the surface to become a paraboloid.

Definition:

A ***vector*** is a quantity which embodies both direction and magnitude. The magnitude is a numerical constant called a ***scalar***.

The most popular application of the vector concept, the one you need in this topic, is that of a **force**. We use arrows of varying length to represent vectors. The arrow specifies the direction and its length indicates the magnitude. Figure 2 (after the next paragraph) shows a slanting vector F with its horizontal component and vertical component. The components themselves are vectors represented by

F_x and by F_y, respectively. Vector F is the sum of the horizontal and vertical vector components. Note that we often place vectors in a coordinate system, and bold letters are customarily used to name vectors.

Suppose an object at the origin was "feeling" a force tugging on it as indicated by vector F. The vector F_x indicates the rightward component of that force and vector F_y shows the upward component of the force. The angle between F and F_x is θ, which is less than $\frac{\pi}{4}$ since the horizontal vector is longer that the vertical one. Therefore more force is exerted to the right than upward by F. Figure 2 shows this configuration.

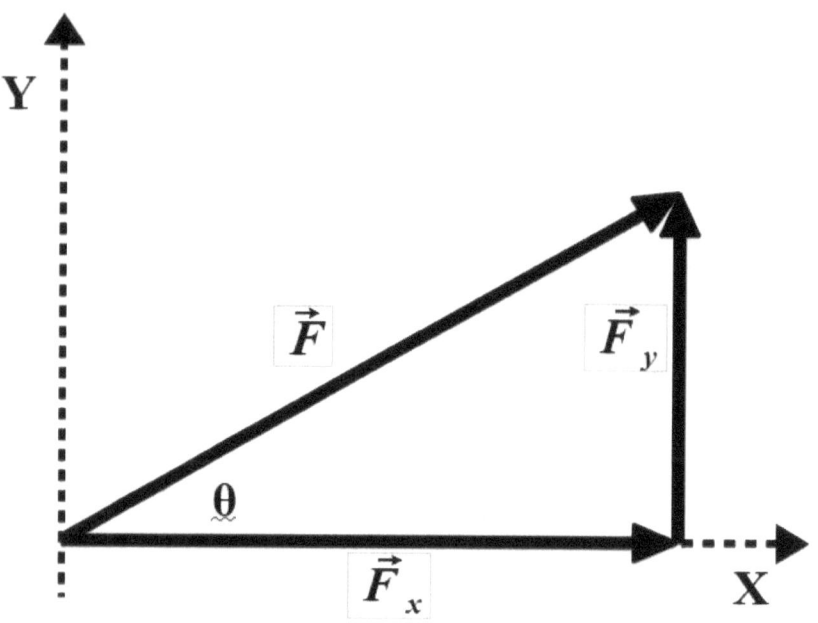

Figure 2

Suppose for example $\theta = \frac{\pi}{6}$ (30⁰). If F has a magnitude of 40 pounds, F_y would have a magnitude of 20 pounds = 40 sin(π/6). Clearly the horizontal vector would have a magnitude of 34.64 pounds = 40 cos(π/6). The vectors form a right triangle. Thus we expect that $\sqrt{(34.64)^2 + (20)^2} = 40$. The diagram shows the accepted vector notation. $\|F\|$ is the notation for the magnitude of F. From the viewpoint of vectors, the vector in the hypotenuse is the ***resultant*** of the vectors F_x and F_y. This is indicated by writing

$$F = F_x + F_y.$$

This does *not* indicate the relationship between their magnitudes. *That* relationship is

$$\|F\| = \sqrt{(\|F_x\|)^2 + (\|F_y\|)^2}.$$

The Rotating Liquid

There are two views of the rotating liquid which we need in order to do a proper analysis as indicated in Figure 3 and Figure 4 showing the top and side views, respectively.

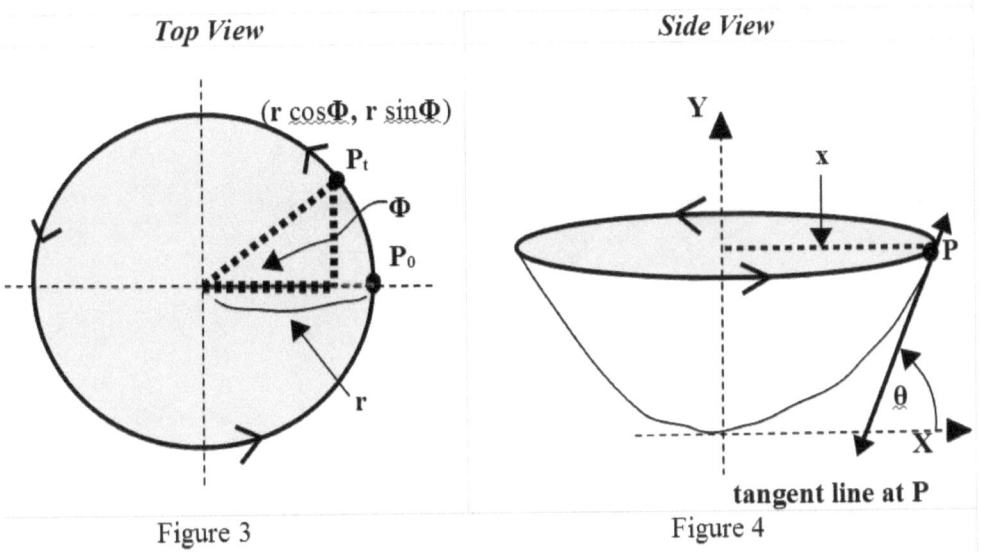

Figure 3

Figure 4

The top view looks directly down on the rotating liquid. Point **P** is on the surface, at the distance **r** from the center of rotation, rotating with the liquid. **P** is subject to two forces: (1) the *centrifugal* force created by the rotation and (2) the force of gravity. We will employ the top view in analyzing the centrifugal force. The side view shows the u-shape of the liquid's surface to which we have drawn a tangent line at point **P**. In this view we use **x** as the radius, where **x** is the x-coordinate of **P** with respect to the axes shown.

The Centrifugal Force

Let's concern ourselves only with the centrifugal force and the top view for the time being. We are assuming the liquid is rotating at a constant rate which we designate by the symbol ω (omega). ω is measured in radians per second and specifies the rotational velocity. In the top view diagram we show point **P** rotating from **P₀** to **Pₜ** moving through Φ radians. Thus $\omega = \frac{d\Phi}{dt}$. Let **R** be the vector from the center to **P**. We can represent the coordinates of **Pₜ** as ($r \cos \Phi$, $r \sin \Phi$). This point specifies the location of the tip end of the vector **F** which is moving around the circle. We imagine this vector rotating with its "tail" at the origin while its "head", attached to **P**, rotates counter-clockwise. We obtain the velocity vector **V** by differentiating the coordinates of **P** with respect to the time, t. Thus

$$\frac{d}{dt}(r \cos \Phi, r \sin \Phi) =$$

$$\left(r(-\sin \Phi) \frac{d\Phi}{dt}, r(\cos \Phi) \frac{d\Phi}{dt} \right)$$

$$= (-r \omega \sin \Phi, r \omega \cos \Phi)$$

We can associate vectors starting at the origin with their "head" coordinates. Thus we have

$$R = (r \cos \Phi, r \sin \Phi)$$
$$V = (-r \omega \sin \Phi, r \omega \cos \Phi)$$

Let's pause and see what we have done here by graphing both of these vectors.

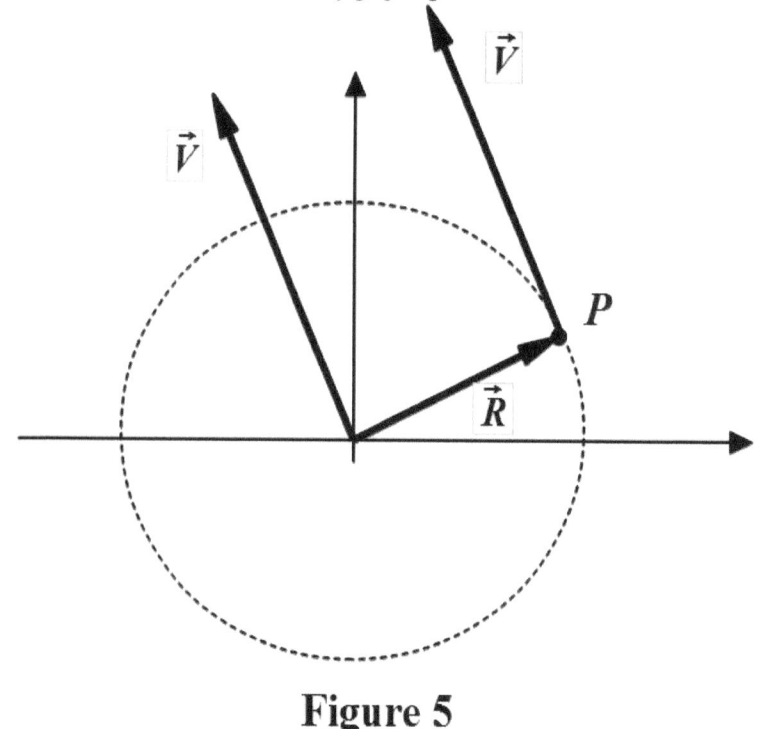

Figure 5

Notice that **V** is drawn in two places and that it is perpendicular to **R**. The slope of **R** is $\frac{r\sin\Phi}{r\cos\Phi} = \tan\Phi$. Looking at **V**, placed at the origin with its head at $(-r\omega\sin\Phi, r\omega\cos\Phi)$, we can get its slope as

$$\frac{r\omega\cos\Phi}{-r\omega\sin\Phi} = -\cot\Phi = \frac{-1}{\tan\Phi}$$

showing that $V \perp R$. We drew the velocity vector with its "toe" at the "head" of **R** as it is an excellent way to show the direction of the movement of **P** as well as the magnitude of the velocity as the length of **V**. Note that the magnitude of **V** is given by

$$\|V\| = \sqrt{(-r\omega\sin\Phi)^2 + (r\omega\cos\Phi)^2} = r\omega.$$

This proves the very reasonable idea that the magnitude of the velocity of a rotating object along a circle of constant radius is proportional to its angular velocity.

Now let's consider the centrifugal force. We use Newton's Law, **F** = m**a**, (force equals mass times acceleration) to get the centrifugal force. First we obtain the acceleration vector, **A**, by differentiating the vector $V = (-r\omega\sin\Phi, r\omega\cos\Phi)$ as follows:

$$A = \frac{d}{dt}(V) = \frac{d}{dt}(-r\omega\sin\Phi, r\omega\cos\Phi)$$

$$A = \left(-r\omega(\cos\Phi)\frac{d\Phi}{dt}, -r\omega(\sin\Phi)\frac{d\Phi}{dt}\right)$$

$$A = (-r\omega^2\cos\Phi, -r\omega^2\sin\Phi)$$

$$A = -\omega^2 R$$

This result, that the acceleration vector is $(-\omega^2)$ times **R**, means that the acceleration is directed inward toward the center. The force vector, according to Newton's Law, can be written as

$$F = mA = -m\omega^2 R$$

The best way to understand why the force is directed inward is to think of swinging a rock on the end of a string around in a circle. The string pulls the rock toward the center. The centrifugal force is the force the rock exerts on the string and is *opposite* to the force given by Newton's Law. The force statement above can be written as a magnitude statement

$$\|F\| = m\omega^2(\|R\|) = mr\omega^2$$

which has the added feature that we do not have to worry about any sign. The centrifugal force is thus proportional to the *square* of the angular velocity.

Just Why Must the Surface be a Paraboloid?

To answer this question we now concentrate on the second view shown earlier, in particular on the tangent line to the curve of the side view of the surface. We draw a magnified version of the second view in Figure 6.

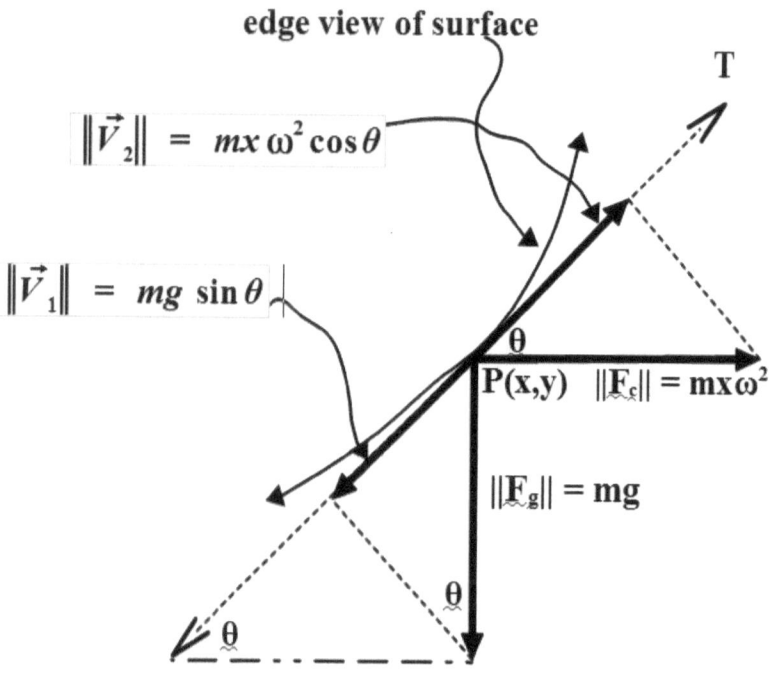

Figure 6

Study carefully the geometry of Figure 6. Consider a small volume of liquid, mass **m**, centered at point P. While rotating at constant velocity, the gravity force and centrifugal forces are in balance, and it is this balance which shapes the surface of the liquid. Line **T** is tangent to the cut away side view of the surface at point **P**. *It is in the direction of this tangent line that the force balance occurs*.

Let θ be the angle of inclination of line **T**. The vertical vector is the gravity force vector and the slanting vector pointing to the lower left is the component of the force of gravity along the tangent line. This vector, V_1 extends from **P** to the foot of the perpendicular dropped to **T** from the "head" of the gravity force vector. By similar triangles, θ appears opposite V_1, so the length (magnitude) of V_1 is **mg** sin θ by using right triangle trigonometry, since the hypotenuse (which is vertical) has length **mg**.

The horizontal vector is the centrifugal force vector which has the magnitude $mx\omega^2$ as seen earlier. It forms the hypotenuse of a triangle formed by dropping a perpendicular from its "head" to line **T**. The vector V_2, is the component of the centrifugal force along the tangent line, and by right triangle trigonometry, its magnitude is **mxω^2 cos θ**.

The balance of forces requires that $\|V_1\| = \|V_2\|$ so that $mg \sin\theta = mx\,\omega^2 \cos\theta$ implies that

$$\frac{\omega^2}{g} x = \tan\theta.$$

Since the tangent of the inclination angle of a line equals the slope of that line, tan θ is also the slope of the curve through **P (x, y)** which is the derivative of the curve at **p**. Thus we have the differential equation

$$\frac{dy}{dx} = \frac{\omega^2}{g} x$$

Separating the variables and integrating, we obtain

$$\int dy = \frac{\omega^2}{g} \int x\, dx$$

$$y = \left(\frac{\omega^2}{2g}\right) x^2 + C$$

Assuming the curve contains the origin at its lowest point (the coordinate system's placement is arbitrary) we have $x = 0$ when $y = 0$. This makes $C = 0$ so the curve of the edge view of the surface is the *parabola*

$$y = \left(\frac{\omega^2}{2g}\right) x^2$$

proving that the surface is indeed a *paraboloid!*

Crossing a Flowing River in Least Time

The Problem

A 500 meter wide river is flowing at 2000 meters per hour. A person rows at an average of 3000 meters per hour (this is the still water rowing rate) and can walk along the shore at 5000 meters per hour. In order to get to the point across the river, directly opposite the starting position, how should the route be planned so as to minimize the total traveling time?

The Solution

As is the usual case with such problems, we need to begin with a meditation period. To help relax and clear the mind for thinking, we draw a diagram. We show two possible

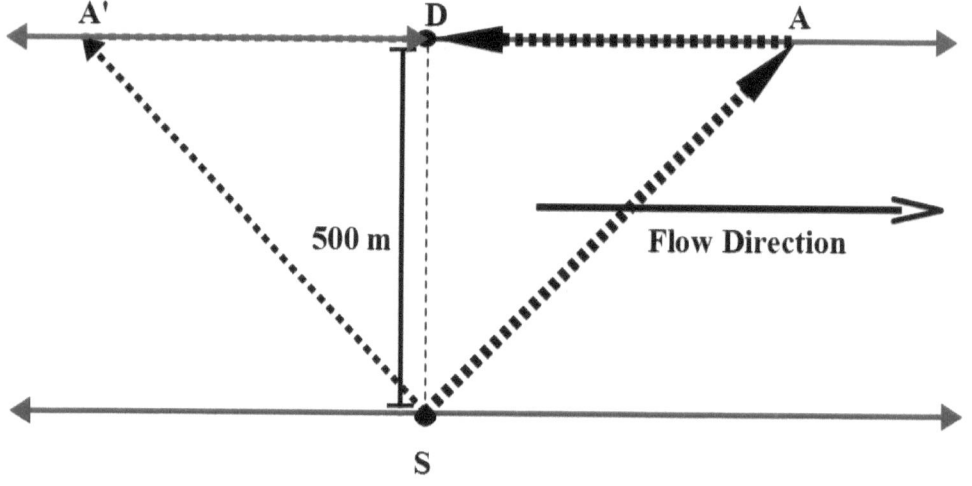

Figure 1

scenarios. From the starting position at **S** the person could take a downstream route to land at point **A** downriver and then walk upriver to point **D**. This route is indicated by the thick dashed arrows from S to A and from A to D. On the other hand, the person could row upstream to land at point **A'** and walk downriver arriving at point **D**, as shown in by the two thinner dashed arrows in Figure 1.

As we think about these possibilities, it must be considered that due to the current, the boat cannot be aimed along either ray **SA** or along ray **SA'**. The boat must be aimed *to the left* instead. Figure 2 shows how this works. The aiming angle is θ as indicated in Figure 2. The horizontal arrow indicates the current which causes the boat to actually move in the direction of the arrow which makes the angle marked Φ with the dashed horizontal line.

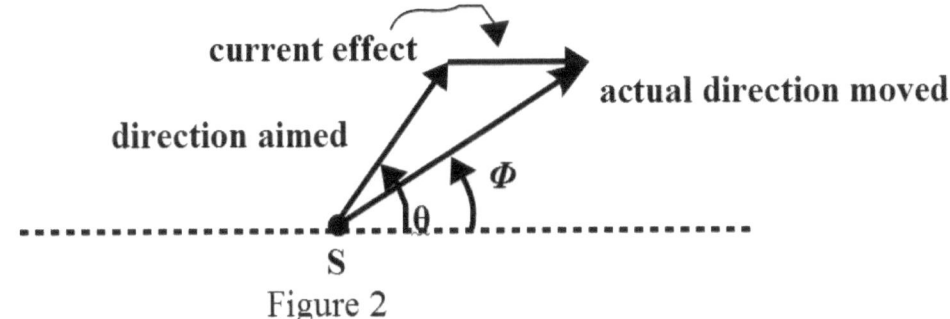
Figure 2

Notice that $0 < \theta < \pi$ since at this point we do not know whether to aim the boat in either the upstream direction $(\frac{\pi}{2} < \theta < \pi)$ or in the downstream direction $(0 < \theta < \frac{\pi}{2})$. Note that $\theta = \frac{\pi}{2}$ is the non-committal (or politician's) direction.

In order to solve this problem we *assume* the downriver direction. We want to develop a function which specifies total traveling time, **T**, in terms of the aiming angle, θ. The underlying idea is the "*time equals distance divided by rate*" concept. There are, of course, two distances to consider – distance on water and distance on land – as the rates of travel are different for each mode. The actual water rate is a composite of the still water rate, 3000 m/h, and the water flow rate, 2000m/h, which is a function of θ. Then we will need to consider how to relate Φ to θ. Figure 3 indicates some important details. Let **R** be the resultant actual rate across the water which takes into account both the flow rate and still water rate.

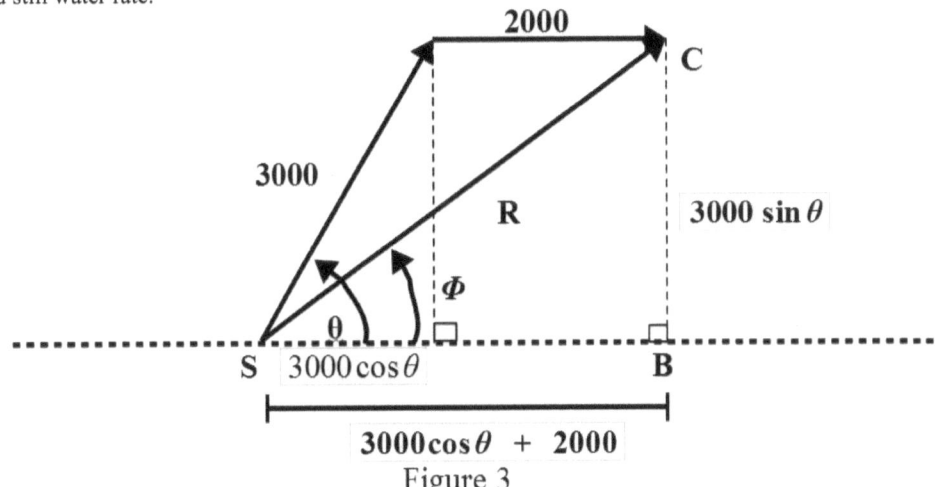
Figure 3

From the Figure 3, using right triangle trigonometry, we see that the horizontal component of **R** is $3000 \cos \theta + 2000$ and its vertical component is $3000 \sin \theta$. Since **R** is the hypotenuse of a triangle whose perpendicular sides have these component lengths, we apply the Pythagorean theorem to obtain

$$R = 1000\sqrt{9\sin^2\theta + (3\cos\theta + 2)^2}$$

Squaring out and simplifying we have

$$R = 1000\sqrt{13 + 12\cos\theta}$$

From right triangle SBC we obtain

$$\cot \Phi = \frac{3000\cos\theta + 2000}{3000\sin\theta}$$

$$\cot \Phi = \frac{3\cos\theta + 2}{3\sin\theta}$$

Below we have reproduced in Figure 4 the original diagram (Figure 1) illustrating the scenario whereby the boat is aimed generally downriver in the direction of the current. We have the function for **T**.

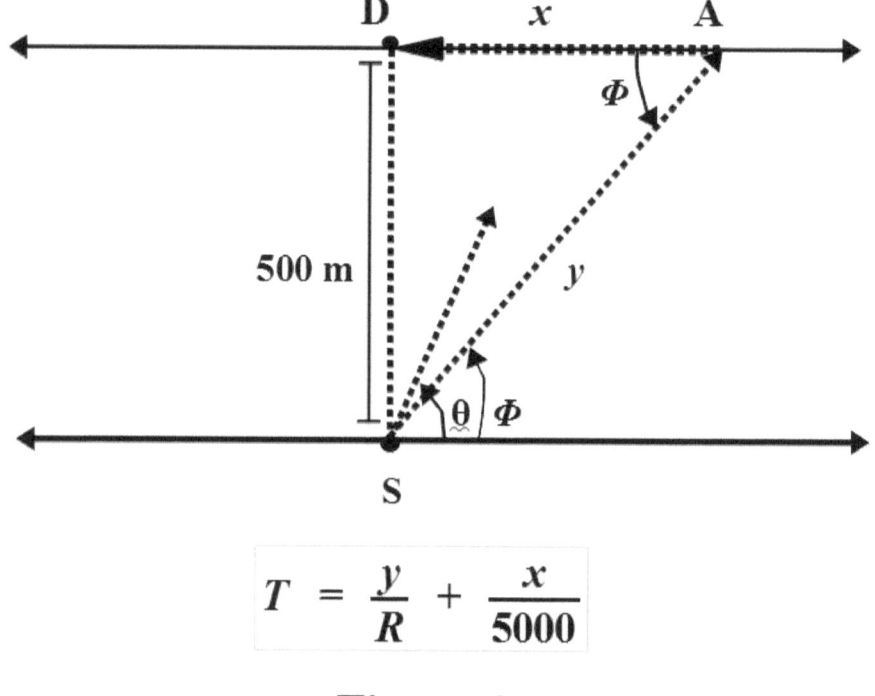

Figure 4

Out of right triangle DAS, $x = 500 \cot \Phi$ and $y = 500 \csc \Phi$. These two relationships and the expression for R on the previous page provide the time function

$$T(\Phi, \theta) = \frac{500 \csc \Phi}{1000\sqrt{13+12\cos\theta}} + \frac{500 \cot \Phi}{5000}$$

which simplifies to

$$T(\Phi, \theta) = \frac{\csc \Phi}{2\sqrt{13+12\cos\theta}} + \frac{\cot \Phi}{10}$$

where $T(\Phi, \theta)$ indicates that we have time as a function of both Φ and θ. Look at the diagram at the bottom of page 2. Right triangle SBC (Figure 3) enables us to write the relationship $\cot \Phi = \frac{3\cos\theta + 2}{3\sin\theta}$ which provides a relationship between Φ and θ. The second term of $T(\Phi, \theta)$ is no problem. To handle the first term, all we need to do is express $\csc \Phi$ in terms of one or more trigonometric functions of θ.

$$\csc \Phi = \sqrt{(\csc^2 \Phi)} = \sqrt{1 + \cot^2 \Phi}$$
$$= \sqrt{1 + \left(\frac{3\cos\theta + 2}{3\sin\theta}\right)^2}$$
$$= \sqrt{\frac{9\sin^2\theta + 9\cos^2\theta + 12\cos\theta + 4}{9\sin^2\theta}}$$
$$\csc \Phi = \frac{\sqrt{13 + 12\cos\theta}}{3\sin\theta}$$

Now we can write the function for the time, T, in terms of θ only.

$$T(\theta) = \frac{1}{6\sin\theta} + \frac{3\cos\theta + 2}{30\sin\theta}$$
$$= \frac{1}{6}\csc\theta + \frac{1}{10}\cot\theta + \frac{1}{15}\csc\theta$$
$$T(\theta) = \frac{7}{30}\csc\theta + \frac{1}{10}\cot\theta$$

Taking the derivative we find

$$T'(\theta) = -\frac{7}{30}\csc\theta\cot\theta - \frac{1}{10}\csc^2\theta$$

We are now looking for critical values in the interval $(0, \pi)$ so we set $T'(\theta) = 0$.

$$\csc\theta(7\cot\theta + 3\csc\theta) = 0$$

implies that

$$\frac{7\cos\theta}{\sin\theta} + \frac{3}{\sin\theta} = 0.$$

The result is that $\cos\theta = -\frac{3}{7}$ so that $\theta = \cos^{-1}\left(-\frac{3}{7}\right)$ which is approximately 115.38°.

If we were to find the second derivative of T(θ) for θ equal to 15.38, the result would be positive, proving that the travel time to the opposite side of the river is at a minimum for this aiming angle. The miracle of the calculus thus tells us that the best path to take to the opposite side of the river is to aim upstream since θ exceeds 90 degrees! Our assumption to aim downriver was thus shown to be wrong.

The Radius of a Black Hole

Introduction

The escape velocity from an astronomical body is the minimum velocity that an object ejected from the surface must have so that it will never fall back to the surface. For the earth this speed is about 25,950 miles per hour or 7 miles per second. For the less massive moon the escape velocity is approximately 1.5 miles per second. For the hugely massive sun the escape velocity from its surface is 1,381,601 miles per hour. Clearly the escape velocity depends on the mass. A black hole is defined as an object which is sufficiently massive and dense so that the escape velocity at its "surface" is equal to the speed of light. Actually a black hole has no real surface since its incredibly intense gravitational force causes its material to collapse to a tiny object termed a singularity. The distance from the singularity where the escape velocity is the speed of light is the so-called radius of the black hole and this boundary region is known as its *event horizon*. Once material slips below this boundary, it cannot escape from the black hole. The photo shows a powerful jet of material being ejected by Galaxy M87 caused by an intense magnetic field created by a black hole at the center of Galaxy M87. See the link http://hubblesite.org/newscenter/archive/releases/2000/20/image/a/ for information about this photo.

Black Hole Jet from the M87 Galaxy – Hubble Photo

In this article we will derive the escape velocity formula. This will require use of the **Work-Energy Theorem** which will also be established. Then the formula for the radius of a black hole, called the *Schwartzschild Radius*, will be developed.

The Work-Energy Theorem

Kinetic energy is energy associated with the movement of an object together with its mass. The formula is $KE = \frac{1}{2}mv^2$ where m is the mass of the object and v is its velocity. The definition of *work* performed on an object is the product of a force and the distance through which the force moves the object. For example, if you lift a book weighing 10 pounds a distance of 3 feet from a table, you have done 30 foot-pounds of work on the book. If you then drop the book 3 feet onto a table, the energy you supplied is transformed into kinetic energy as the book lands on the table, having achieved a certain velocity contributed by gravity. The work done by you is equal to the kinetic energy gained by the book as it hits the table. This equality is stated by the Work-Energy Theorem whose formal statement is below.

Let F be the resultant of force in the direction of movement of an object along an axis from point A where x = a to point B where x = b. Let v(x) be the velocity of the object at the position where its coordinate (position on the axis) is x. Then the work, W, done by F in moving the object from point A to point B is given by

$$W_A^B = \int_a^b F(x)\,dx = \frac{1}{2}m(v(b))^2 - \frac{1}{2}m(v(a))^2$$

or

$$W_A^B = \Delta KE_A^B$$

Proof: According to Newton's formula relating force, mass, and acceleration, $F = ma$. The integral appears in the work-energy expression since the force varies from a to b. We re-define the acceleration a by use of the Chain Rule as follows.

$$a = \frac{dv}{dt} = \frac{dv}{dx} * \frac{dx}{dt} = v\frac{dv}{dx}$$

This allows us to perform the integration to calculate the work.

$$W_A^B = \int_a^b F\,dx = \int_a^b ma\,dx =$$

$$\int_a^b mv\frac{dv}{dx}dx = \int_{v(a)}^{v(b)} mv\,dv$$

$$W_A^B = \frac{1}{2}mv^2 \Big|_{v(a)}^{v(b)}$$

$$W_A^B = \frac{1}{2}m(v(b))^2 - \frac{1}{2}m(v(a))^2$$
$$= \Delta KE_A^B$$

Escape Velocity Formula

To obtain a formula for the escape velocity we imagine "lifting" an object from the surface of a planet to an infinite distance and then dropping it there. Then the planet's gravity tugs on it, minutely at first, then more greatly as it falls back to the surface. As it strikes the surface it has at that time attained a velocity which, if the direction of movement were reversed, would be the escape velocity.

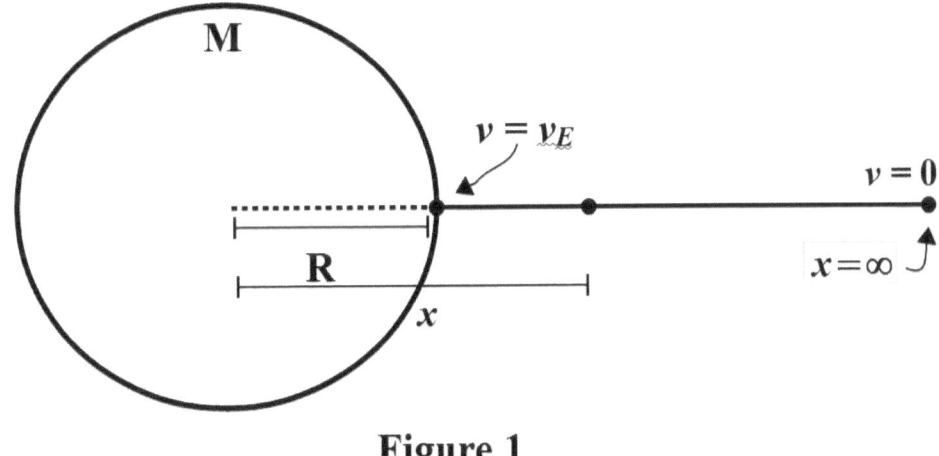

Figure 1

Figure 1 shows this scenario. The object is placed way to the right infinitely far from an object of mass **M**, radius **R** at an initial velocity of 0. It falls to the left eventually reaching the escape velocity v_E when it contacts the surface of the object at the distance of **R** from the object's center of mass. Let m be the mass of the falling object. When applying the Work-Energy Theorem we need the gravitational force formula $F = \dfrac{G\, m\, M}{x^2}$ where F is the gravity force, x is the distance between the centers of mass, and G is the universal gravitation constant. At an infinite distance, the point of release, the kinetic energy is zero, but when it reaches the surface, gravity has supplied the kinetic energy $\frac{1}{2} m v_E^2$. Therefore by the Work Energy Theorem we have

$$W_R^\infty = \int_R^\infty \left(\frac{G\,M\,m}{x^2}\right) dx = \lim_{t \to \infty} \int_R^t \left(\frac{G\,M\,m}{x^2}\right) dx$$

$$= \lim_{t \to \infty} \left(\frac{-G\,M\,m}{x}\bigg|_R^t\right) = (G\,M\,m) \lim_{t \to \infty} \left(\frac{1}{R} - \frac{1}{t}\right)$$

$$W_R^\infty = \frac{G\,M\,m}{R} = \frac{1}{2} m v_E^2$$

Canceling the mass, m and solving for v_E we obtain the escape velocity formula.

$$v_E = \sqrt{\frac{2GM}{R}} \tag{1}$$

The Schwartzschild Radius

For a black hole we replace v_E with c, the speed of light. Solving (1) for $R = R_s$, the Schwartzschild Radius, we have

$$R_s = \frac{2GM}{v_E^2}.$$

For the sun, R_s = 1.9 miles. If the earth were compressed to a sphere of a radius of about one centimeter, it would become a black hole. Our Milky Way galaxy contains a 3 million solar mass black hole while the Andromeda Galaxy contains a 30 million solar mass black hole at its center. The largest black hole detected so far has a mass of 21 billion suns. This makes its size comparable to an entire galaxy.

The Two Cannon Projectile Scenario

PROBLEM: Suppose two cannons are aimed toward one another, one atop a cliff, the other at a lower level. *Simultaneously*, each fires a projectile at initial velocities of not necessarily equal magnitude. Prove that the two projectiles must collide (at a point possibly below the lower cannon).

PROOF:

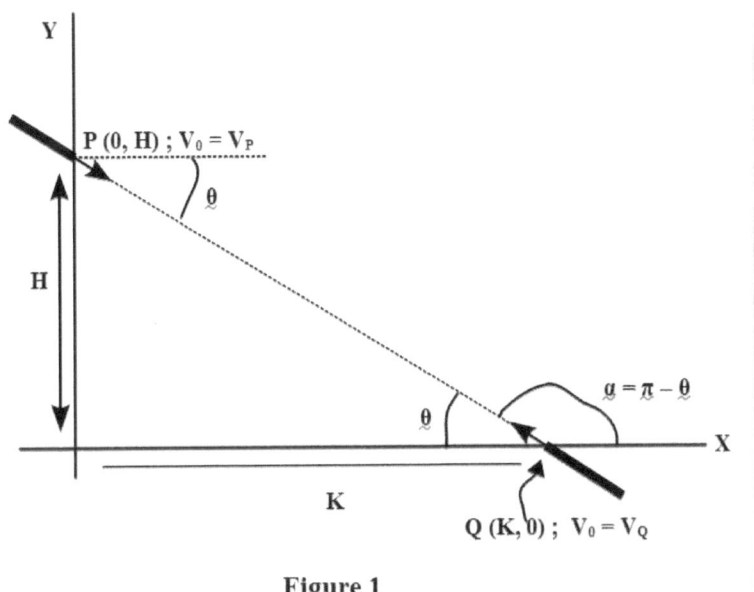

Figure 1

Place one cannon at point P (0, H) as shown in Figure 1 with initial muzzle velocity of V_P and the other at point Q(K, 0) with initial muzzle velocity of V_Q. Then at time $t = 0$ we have the following initial conditions:

At P:	At Q:
$X_P = 0$	$X_Q = K$
$Y_P = H$	$Y_Q = 0$
$V_X = V_P \cos \theta$	$V_X = V_Q \cos(\pi - \theta)$
	$= -V_Q \cos \theta$
$V_Y = -V_P \sin \theta$	$V_Y = V_Q \sin(\pi - \theta)$
	$= V_Q \sin \theta$
$a_X = 0$	$a_X = 0$

27

$$a_y = -g \qquad a_y = -g$$

For the path of the projectile from the cannon at point P we develop parametric equations expressing X and Y in terms of t.

$$a_x = \frac{dV_x}{dt} = 0, \; V_x = C_1, \; V_x = V_P \cos\theta$$

$$a_y = \frac{dV_y}{dt} = g, \; V_y = -gt + C_2, \; V_y = -gt - V_P \sin\theta$$

$$V_x = \frac{dX_P}{dt} = V_P \cos\theta \quad X_P = V_P t \cos\theta$$

$$V_y = \frac{dV_P}{dt} = -gt - V_P \sin\theta, \; Y_P = -\frac{1}{2}gt^2 - V_P t \sin\theta + H$$

The path of the projectile shot from the cannon at point P is given by the system:

$$\left\{ \begin{array}{l} X_P = V_P t \cos\theta \\ Y_P = -\frac{1}{2} g t^2 - V_P t \sin\theta + H \end{array} \right\} \qquad (1)$$

Similarly we develop parametric equations for the path of the projectile shot from the cannon at point Q.

$$a_x = \frac{dV_x}{dt} = 0, \; V_x = C_3, \; V_x = V_Q \cos\theta$$

$$a_y = \frac{dV_y}{dt} = g, \; V_y = -gt + C_4, \; V_y = -gt - V_Q \sin\theta$$

$$V_x = \frac{dX_Q}{dt} = -V_Q \cos\theta, \; X_Q = -V_Q t \cos\theta + K$$

$$V_y = \frac{dV_Q}{dt} = -gt + V_Q \sin\theta, \; Y_Q = -\frac{1}{2} g t^2 + V_Q t \sin\theta$$

Thus the path of the projectile shot from the cannon at point Q is given by the system:

$$\left\{ \begin{array}{l} X_Q = -V_Q t \cos\theta + K \\ Y_Q = -\frac{1}{2} g t^2 + V_Q t \sin\theta \end{array} \right\} \qquad (2)$$

To prove that the two projectiles collide we must show that they are at the same point (X, Y) for the *same* value of t. We begin by setting the X-expressions equal in systems (1) and (2) to obtain the time at which each projectile has the same X-coordinate. Thus if $X_P = X_Q$ we have

$$V_P t \cos\theta = K - V_Q t \cos\theta$$

$$t = \frac{K}{(V_P + V_Q)\cos\theta}$$

We now take the Y-equations in systems (1) and (2) and substitute this value of t. Starting with system (1) we have

$$Y_P = -\frac{1}{2}g\left(\frac{K}{(V_P+V_Q)\cos\theta}\right)^2 - V_P\left(\frac{K}{(V_P+V_Q)\cos\theta}\right)\sin\theta + H$$

$$= \frac{-gK^2}{2(V_P+V_Q)^2\cos^2\theta} - \left(\frac{V_P K}{V_P+V_Q}\right)\tan\theta + H$$

Since $\tan\theta = \frac{H}{K}$, we find that

$$Y_P = \frac{-gK^2}{2(V_P+V_Q)^2\cos^2\theta} - \left(\frac{V_P}{V_P+V_Q}\right)H + H$$

$$= \frac{-gK^2}{2(V_P+V_Q)^2\cos^2\theta} + H\left(\frac{V_P+V_Q-V_P}{V_P+V_Q}\right)$$

$$Y_P = \frac{-gK^2}{2(V_P+V_Q)^2\cos^2\theta} + H\left(\frac{V_Q}{V_P+V_Q}\right)$$

Now from system (2) we have

$$Y_Q = -\frac{1}{2}g\left(\frac{K}{(V_P+V_Q)\cos\theta}\right)^2 + V_Q\left(\frac{K}{(V_P+V_Q)\cos\theta}\right)\sin\theta$$

$$= -\frac{1}{2}g\left(\frac{K}{(V_P+V_Q)\cos\theta}\right)^2 + \left(\frac{V_Q}{(V_P+V_Q)}\right)K\tan\theta$$

$$= \frac{-gK^2}{2(V_P+V_Q)^2\cos^2\theta} + H\left(\frac{V_Q}{V_P+V_Q}\right)$$

$$Y_Q = Y_P$$

Therefore both projectiles are at the same position at the same time.

We note here that if we want the collision to occur above the level of point Q we would require that

$$\frac{-g K^2}{2(V_P + V_Q)^2 \cos^2 \theta} + H\left(\frac{V_Q}{V_P + V_Q}\right) > 0$$

which is equivalent to the condition

$$V_Q(V_P + V_Q) > \frac{g}{2}\left(\frac{H^2 + K^2}{H}\right).$$

Modeling Average Temperatures with Complex Numbers

Introduction

Unlike the majority of articles in this book, this item does not employ calculus. Nevertheless it is an unusual adventure in mathematics. The recursive sequence which is the star of this development was discovered by the author while doing a mathematical doodle during a particularly boring faculty meeting at a high school. His curiosity sparked by the odd nature of the recursive sequence defined by (2) lasted for three decades until a breakthrough was made in discovering a formula whose graph contains terms of sequence (2). An early attempt was made by analyzing the simpler sequence defined by (3). It turned out that (3) unexpectedly led to the process shown here in describing climate. That odyssey appears in the next non-calculus article in this book entitled "A Curious Recursive Sequence". That adventure leads to an interesting surprise.

The purpose of this discussion is the presentation of a process whereby a function of the form $y = T(x)$ can be devised which pairs the day of the year whose number is x, $1 \leq x \leq 365$, with the average temperature, $T(x)$, to be expected on day x. The function is not predictive – rather, it is descriptive of the climate. The general form of the formula to be developed is

$$Y = T(x, T_1, T_2, D, A),$$

in which the parameters T_1, T_2, D, and A are constants suitable for the location where the temperatures are to be modeled. We also present a description of how this formula arose.

A Recursion Formula

You are likely to be familiar with the Fibonacci sequence whose recursion (generating) formula is

$$a_n = a_{n-1} + a_{n-2} \tag{1}$$

Taking this formula with the integer n being 3 or more and $a_1 = 0$, $a_2 = 1$, we obtain the sequence 0, 1, 1, 2, 3, 5, 8, 13,

The thought occurred, as noted in the introduction, that an interesting recursion sequence might be generated by use of the formula

$$a_n = a_{n-1} - a_{n-3}, a_1 = 1, \ a_2 = 1, \ a_3 = 1, \ n > 3. \tag{2}$$

Some terms of this sequence are

1, 1, 1, 0, −1, −2, −2, −1, 1, 3, 1, 4, 3, 0, −4, −7, −3, 4, 11, 14, 10, −1, ⋯

A simpler sequence is defined in (3) which is the basis for this article.

$$a_n = a_{n-1} - a_{n-2}, \ a_1 = 1, \ a_2 = 1, \ n > 2 \tag{3}$$

The sequence generated by (3) is

$$1, 1, 0, -1, -1, 0, 1, \quad 1, 0, -1, -1, 0, 1, 1, \cdots$$

Graphing this sequence, pairing the first term with 1, we see the graph in Figure 1.

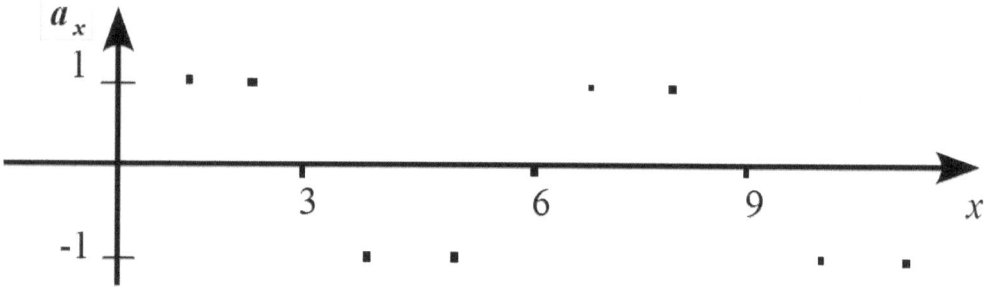

Figure 1

The graph looks sinusoidal with a period of 6 units with an amplitude somewhat greater than 1. Having done a high school science project on average temperatures, the author saw a connection after imagining a smooth curve passing through the points in Figure 1. If the year is divided into sixths, look at the data in Figure 2 showing the average temperature expected on the date noted. See what happens if 10 is subtracted from these temperatures? The numbers in the bottom row

Date	4/25	6/24	8/24	10/24	12/24	2/23
Average	10	21	21	11.6	-0.6	-0.6
Average - 10	0	11	11	1.6	-10.6	-10.6

Figure 2

appear to approximate a sequence generated by (3) but with $a_1 = 0$, $a_2 = 11$.

Modeling Average Temperatures in General

Let A be the average annual temperature for a given location, and T_1, T_2, T_3, T_4, T_5, and T_6 be the average temperature at times one-sixth of a year apart. Then

$$\frac{T_1 + T_2 + T_3 + T_4 + T_5 + T_6}{6} = A$$

and we propose, based on observations made in diverse locations, that

$$(T_2-A)-(T_1-A)+A = T_2-T_1+A \approx T_3.$$

Figure 2 contains the Boston data. The bottom row shows the T_n temperatures minus $A = 10$, and it is evident that $11 - 0 = 11 = T_3$. Also $11 - 11 = 0$ which is approximately $1.6 = T_4 - 10$ and so on. In general we propose that

$$T_n = (T_{n-1}-A)-(T_{n-2}-A)+A = T_{n-1}-T_{n-2}+A$$

where $n = 3, 4, 5,$ and 6, is a fundamental principle of climate governing the changing of the average daily temperature with the seasons. Here is what this means. If you take a set of expected average temperatures on *any* six equally spaced days and subtract from them the annual mean temperature, the six numbers obtained satisfy (3)! This has been successfully verified for locations as varied as Boston, Massachusetts and Fort Lauderdale, Florida. Strong evidence exists that this property is true of the average temperatures in Anchorage, Alaska.

We begin with a general sequence involving p and q which satisfies (3). Taking $p = a_1$ and $+q = a_2$ we have the sequence

$$p, q, q-p, -p, -q, p-q, p, q, \cdots \qquad (4)$$

We now seek a formula for a_n strictly in terms of p and q. (Note that in Figure 2, $p = 0$ and $q = 1$.) The process involves use of geometric sequences. Let $1, S, S^2, S^3, \cdots$ be any geometric sequence satisfying formula (3). Then we have

$$S^n = S^{n-1} - S^{n-2}$$

or

$$S^n - S^{n-1} + S^{n-2} = 0$$

Multiplying through by S^{2-n} we get

$$S^2 - S + 1 = 0$$

which, by the quadratic formula, has the solutions $S = \dfrac{1 \pm i\sqrt{3}}{2}$. You should notice that these solutions are the two complex cube roots of -1. To make the notation easier, for the rest of this discussion let $\alpha = \dfrac{1+i\sqrt{3}}{2}$ and $\beta = \dfrac{1-i\sqrt{3}}{2}$. Thus we have two sequences of numbers satisfying (3) as given by

$$1, \alpha, \alpha^2, \alpha^3, \cdots \text{ and } 1, \beta, \beta^2, \beta^3, \cdots$$

Let c and d be any two constants. It is easily established that if any sequence satisfies (3), any constant multiple of that sequence also satisfies (3). Also, if the sequences $\{a_n\}$ and $\{b_n\}$ satisfy (3), then the sequence $\{a_n + b_n\}$ satisfies (3). Thus the most general sequence satisfying (3) is the sequence

$$(c+d), (c\alpha+d\beta), (c\alpha^2+d\beta^2), (c\alpha^3+d\beta^3), \cdots \tag{5}$$

Looking back to (4) we want $p = c + d$, and $q = c\alpha + d\beta$. Solving this set of simultaneous equations for c and d we obtain

$$c = \frac{q-p\beta}{\alpha-\beta}, \quad d = \frac{p\alpha-q}{\alpha-\beta} \tag{6}$$

Since sequence (5) is the general solution of the recursion formula (3) we have

$$a_1 = c+d, \; a_2 = c\alpha+d\beta, \; a_3 = c\alpha^2+d\beta^2, \cdots,$$

so that

$$a_n = c\alpha^{n-1} + c\beta^{n-1} \tag{7}$$

is the desired expression for the nth term strictly in terms of p and q. Substituting the values found in (6) into (7) we have

$$a_n = \left(\frac{q-p\beta}{\alpha-\beta}\right)\alpha^{n-1} + \left(\frac{p\alpha-q}{\alpha-\beta}\right)\beta^{n-1} \tag{8}$$

Substituting $\alpha = \frac{1+i\sqrt{3}}{2}$ and $\beta = \frac{1-i\sqrt{3}}{2}$ into (8) and doing a considerable amount of algebraic simplification, equation (8) becomes

$$a_n = \left(\frac{3p+(p-2q)i\sqrt{3}}{6}\right)\left(\frac{1+i\sqrt{3}}{2}\right)^{n-1} + \left(\frac{3p-(p-2q)i\sqrt{3}}{6}\right)\left(\frac{1-i\sqrt{3}}{2}\right)^{n-1} \tag{9}$$

Formula (7) is structured so that $a_1 = p$, $a_2 = q$, $a_3 = q - p$, $a_4 = -p$, $a_5 = -q$, $a_6 = p - q$, $a_7 = p$, ... Thus (9) has a period of 6. Considering n to be a real number instead of an integer, Figure 3 shows the graph of the function in (9) considering n to be continuous with the rectangular dots showing the points where n is an integer.

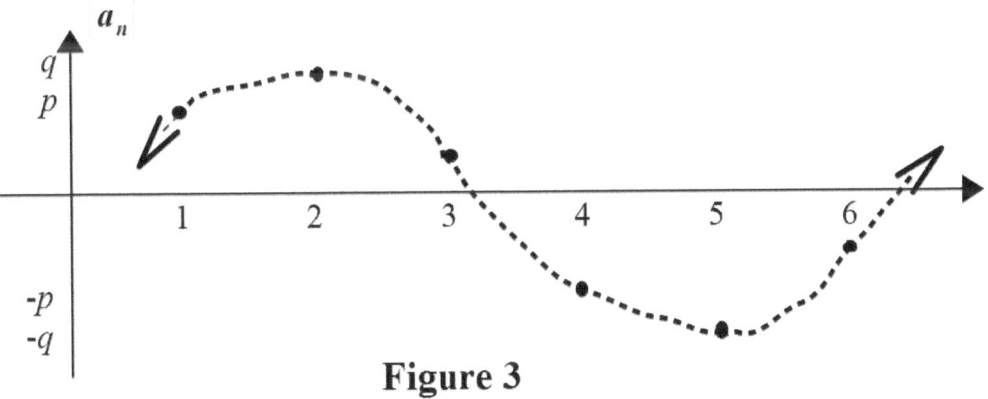

Figure 3

The formula for the average temperature on any day is developed from equation (9). Figure 4 indicates a typical plot of average temperature, T(x) as a function of the day number, x. Think of the

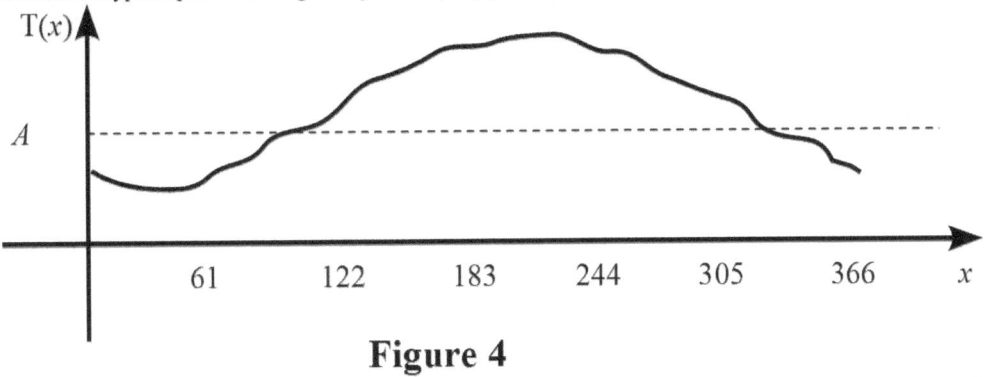

Figure 4

graph in Figure 4 as an annual plot of daily temperatures averaged over, say, 20 years. As the number of years increases, the smoothness of the curve increases, and eventually, the average temperature curve for any given location looks like a portion of a sine curve. Then we can say the function $y = T(x)$ describes the climate with respect to the daily *expected* average temperature as a function of the date. Note that A is the average annual temperature, the average of all the y-coordinates shown in the graph.

We are going to modify equation (9) so that its graph becomes like that shown in Figure 4. The horizontal axis in Figure 4 is divided into 6 equal intervals with the year taken as having 366 days for convenience. First we add A to (9) to lift its graph and replace n with $z/61$ to stretch the graph horizontally, giving it a period of $366 = 6(61)$. Equation (10) shows the result of this modification (best viewed in landscape mode).

$$T(z) = \left(\frac{3p+(p-2q)i\sqrt{3}}{6}\right)\left(\frac{1+i\sqrt{3}}{2}\right)^{\frac{(z-61)}{61}} + \left(\frac{3p-(p-2q)i\sqrt{3}}{6}\right)\left(\frac{1-i\sqrt{3}}{2}\right)^{\frac{(z-61)}{61}} + A \qquad (10)$$

Let T_1 and T_2 be the average temperature observed on two days which are one-sixth (61 days) apart with T_1 being the earlier. Let D be number of days from the end of the year when T_1 is observed. This makes T_2 the temperature occurring $D - 61$ days from the end of the year. As (10) is now written, $z = 61$ corresponds to $n = 1$ in (9) and thus $T(61) = p + A$. If $z = 122$, this corresponds to $n = 2$ in (9) and thus $T(122) = q + A$. Similarly, $T(183) = (q - p) + A$, etc. The sequence (4) in the light of the $T(z)$ function is

$$T(61) - A, \quad T(122) - A, \quad T(183) - A, \quad T(244) - A, \quad T(305) - A, \quad T(366) - A, \ldots$$

Therefore we must replace p and q, respectively, with $T_1 - A$ and $T_2 - A$.

We need to adjust z by replacing it with a function of x so that, according to the selection of T_1 and T_2, $T(x) = $ the temperature expected on day number x, where $1 \le x \le 366$. Thus, for example, $T(5)$ would be the temperature expected on January 5. This amounts to a horizontal shift in the graph in Figure (3) along with the other modifications. We must involve the number D in doing this. Clearly $z = 61$ corresponds to the first observed temperature and $x = 366 - D$ corresponds to the first observed average temperature, T_1. Note that $z - 61 = 0$ and $x + D - 366 = 0$ so the correct relationship is $z - 61 = x + D - 366$ and

$$\frac{z-61}{61} = \frac{x+D-366}{61} = \left(\frac{x+D}{61}\right) - 6.$$

Because $\alpha = \frac{1+i\sqrt{3}}{2}$ and $\beta = \frac{1-i\sqrt{3}}{2}$ are cube roots of -1, it is easy to see that $\alpha^{t-6} = \alpha^t$ and $\beta^{t-6} = \beta^t$. Therefore we drop the -6 and replace $\frac{z-61}{61}$ with $\frac{x+D}{61}$ in equation (10). After all modifications we have the result in Figure 5.

$$T(x) = \left(\frac{3(T_1-A)+(T_1-2T_2+A)i\sqrt{3}}{6}\right)\left(\frac{1+i\sqrt{3}}{2}\right)^{(x+D)/61} +$$
$$\left(\frac{3(T_1-A)-(T_1-2T_2+A)i\sqrt{3}}{6}\right)\left(\frac{1-i\sqrt{3}}{2}\right)^{(x+D)/61} + A \qquad (11)$$

where $T_1 = $ is the average temperature observed D days before the end of the year
$T_2 = $ is the average temperature observed 61 days after the day T_1 is observed
$A = $ the average annual (mean) temperature for the entire year
$i = $ the square root of -1
$x = $ a day number from 1 to 365
$T(x) = $ the average temperature expected on day number x

Figure 5

The Trigonometric Form of the General Average Temperature Function

Equation (11) is not at all convenient for doing calculations. Using DeMoivre's formula we can derive a trigonometric form. The resulting function will not involve $\sqrt{-1}=i$. Using polar form it is evident that

$$\alpha = \frac{1+i\sqrt{3}}{2} = \cos\left(\frac{\pi}{3}\right)+i\sin\left(\frac{\pi}{3}\right) \quad \text{and} \quad \beta = \frac{1-i\sqrt{3}}{2} = \cos\left(\frac{\pi}{3}\right)-i\sin\left(\frac{\pi}{3}\right).$$

DeMoivre's formula tells us that

$$(\cos\theta \pm i\sin\theta)^p = \cos(p\theta) \pm i\sin(p\theta).$$

Replacing θ with $\frac{\pi}{3}$, α and β with their polar forms, and then doing an enormous amount of simplifying, equation (11) in Figure 5 becomes

$$T(x) = (T_1 - A)\cos\left(\frac{\pi(x+D)}{183}\right) + \left(\frac{(2T_2 - T_1 - A)\sqrt{3}}{3}\right)\sin\left(\frac{\pi(x+D)}{183}\right) + A \tag{12}$$

This is the climate describing average temperature function in a practical form.

Actual Data – How Well Does the Formula Describe Climate?

For the Boston metropolitan area the June 24 temperature expected, on average, is 21°C on a date which is 190 days from the end of the year. Sixty one days later the August temperature is also expected to be 21°C. The annual mean temperature for Boston is 10°C. Taking formula (12) and substituting $T_1 = T_2 = 21$ and $A = 10$ we get

$$T(x) = 11\cos\left(\frac{\pi(x+190)}{183}\right) + \left(\frac{11\sqrt{3}}{3}\right)\sin\left(\frac{\pi(x+190)}{183}\right) + 10$$

This formula has been tested against all of the average daily temperatures based on a record including at least the past 140 years. 94% of the computed average temperatures were within 1.5 degrees of the published daily averages. Interestingly, all of the computed temperatures from September 14 to April 24 were below the actual data. Boston area residents know quite well that the temperatures reported at Logan Airport are higher in the winter than the temperatures in the suburbs due to the heat absorption of the runways. Thus the formula may actually be eliminating a local environmental effect.

For Fort Lauderdale, on January 15, 351 days from the end of the year, the expected temperature is $T_1 = 20.11°$ C and 61 days later, on March 17, the expected average temperature is $T_2 = 21.33°$ C. The annual mean temperature is $A = 24.11°C$. For Fort Lauderdale the function is

$$T(x) = -4\cos\left(\frac{\pi(x+351)}{183}\right) - 0.9\sin\left(\frac{\pi(x+190)}{183}\right) + 24.11$$

A Curious Recursive Sequence and a More Curious n^{th} Term Formula

Introduction

The subject of this article is the recursive sequence, a few of whose terms are

$$1, 1, 1, 0, -1, -2, -2, -1, \quad 1, 3, 4, 3, 0, -4, -7, \cdots \qquad (1)$$

It was discovered by the author during a particularly boring and irrelevant faculty meeting in an attempt to keep the mind busy. In effect, it was a mathematical doodle. The recursive formula for the above sequence is

$$t_{n+1} = t_n - t_{n-2}, n \geq 3,$$

where $t_1 = t_2 = t_3 = 1$. We can run the sequence backwards by rewriting this formula as follows:

$$t_{n-2} = t_n - t_{n+1}$$
$$t_{n-1} = t_{n+1} - t_{n+2}$$

The second formula is convenient for this purpose. Let $n = 1$. Then we have

$$t_0 = t_2 - t_3$$

Thus $t_0 = 1 - 1 = 0$. Similarly, $t_{-1} = t_1 - t_2 = 1 - 1 = 0$, $t_{-2} = t_0 - t_1 = 0 - 1 = -1$. This extends the sequence back three terms, so we how have

$$-1, 0, 0, 1, 1, 1, 0, -1, -2, -2, \quad -1, 1, 3, 4, 3, 0, -4, -7, \cdots.$$

When extended further backward and forward, this sequence exhibits two different behavior patterns. We now list some more terms which will be referred to as (2).

…,86,-65,49,-37,28,-21,16,-12,9,-7,5,-4,3,-2,2,-1,1,-1,0,-1,0,0,1,1,1,0,-1,-2,-2,-1,1,3,4,3,0,-4,-7,-7,-3,4,11,14,10,-1,-15,-25,-24,-9,-16,40,49,33,-7,-56,-89,-82,-26,63, … (2)

After inspecting this sequence, a graph of these terms clearly indicates a dual behavior. Pairing the three one's in the first sequence list with, respectively, $x=1$, $x=2$, and $x=3$, we can create the following graph, created with the TI89 Graphing Calculator. The program plotted a few pixels for each point for better legibility.

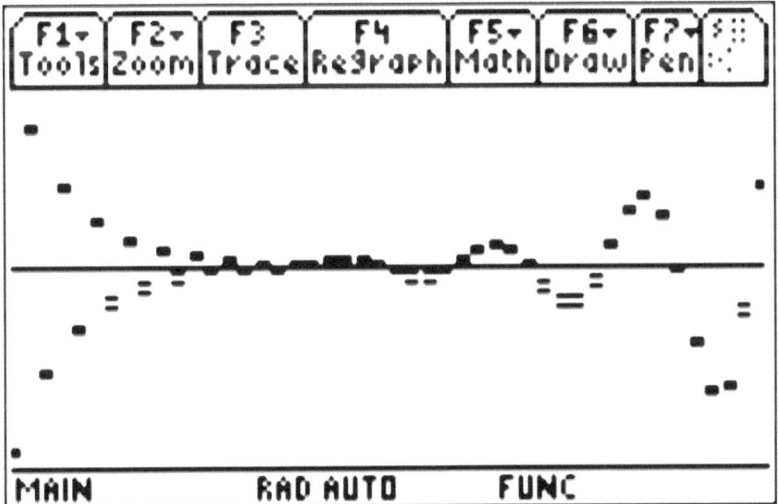

Figure 1

In view of the rather strange pattern, one wonders how an nth term function might appear which would have as its domain the set of integers and the recursive term values in (2) in its range. The determination of this function and its domain extension to the set of real numbers is the subject of this article. The graph of the nth term function over the real number set is surprising.

The Fibonacci Sequence

Since readers are likely to be familiar with the Fibonacci Sequence, we will motivate some of the ideas in the discussion of (2) by analyzing this famous sequence first. The sequence

$$1,1,2,3,5,8,13,21,34,55,89,144, \ldots$$

has as its generating formula

$$F_{n+2} = F_{n+1} + F_n, \; n \geq 1$$

where $F_1 = 1$ and $F_2 = 1$. The so-called characteristic polynomial equation is formed as follows.

$$x^{n+2} = x^{n+1} + x^n$$
or
$$x^2 - x - 1 = 0$$

Solving this equation yields two roots:

$$\Phi = \frac{1+\sqrt{5}}{2}, \quad \alpha = \frac{1-\sqrt{5}}{2}. \tag{3}$$

Note that the number Φ is referred to as the golden ratio. Taking Φ first, we see the following powers of Φ.

$$\begin{aligned}
\Phi^2 &= & \Phi+1 \\
\Phi^3 &= \Phi^2+\Phi = & 2\Phi+1 \\
\Phi^4 &= \Phi^3+\Phi^2 = & 3\Phi+2 \\
\Phi^5 &= \Phi^4+\Phi^3 = & 5\Phi+3 \\
\Phi^6 &= \Phi^5+\Phi^4 = & 8\Phi+5 \\
&\cdots&
\end{aligned}$$

In the right column the coefficients of Φ and the constant terms are themselves Fibonacci numbers. We can write the same table for powers of α. A careful examination of these equations for powers of Φ and α leads to

$$\Phi^n = (F_n)(\Phi) + F_{n-1}$$
$$\alpha^n = (F_n)(\alpha) + F_{n-1}$$

Subtracting the second equation from the first yields

$$(F_n)(\Phi - \alpha) = (\Phi^n - \alpha^n)$$

which leads immediately to the nth term formula

$$(F_n) = \frac{(\Phi^n - \alpha^n)}{(\Phi - \alpha)}.$$

Using the values in (3) we get the final version for F_n

$$F_n = \frac{1}{\sqrt{5}}\left(\left(\frac{1+\sqrt{5}}{2}\right)^n - \left(\frac{1-\sqrt{5}}{2}\right)^n\right).$$

The Curious Sequence

We are now ready to turn our attention to the sequence in (1) for the purpose of determining the nth term formula, taking $t_1 = t_2 = t_3 = 1$. From the rule for generating new terms,

$$t_{n+1} = t_n - t_{n-2}, \quad n \geq 3,$$

we add 2 to each subscript to obtain

$$t_{n+3}=t_{n+2}-t_n$$

Forming the characteristic polynomial equation, we have

$$x^{n+3}=x^{n+2}-x^n$$

or

$$x^3-x^2+1=0$$

This cubic equation has three roots which will concern us later. For now, let α represent any of the three roots. The following equations are a result of the relation $x^3 = x^2 - 1$.

$$
\begin{aligned}
\alpha^3 &= \alpha^2+0-1 \\
\alpha^4 &= \alpha^3-\alpha = \alpha^2-\alpha-1 \\
\alpha^5 &= \alpha^4-\alpha^2 = (\alpha^2-\alpha-1)-\alpha^2 = 0-\alpha-1 \\
\alpha^6 &= \alpha^5-\alpha^3 = (\alpha-1)-(\alpha^2-1) = -\alpha^2-\alpha+0 \\
\alpha^7 &= \alpha^6-\alpha^4 = (-\alpha^2-\alpha)-(\alpha^2-\alpha-1) = -2\alpha^2+0+1 \\
\alpha^8 &= \alpha^7-\alpha^5 = (-2\alpha^2+1)-(-\alpha-1) = -2\alpha^2+\alpha+2 \\
\alpha^9 &= \alpha^8-\alpha^6 = (-2\alpha^2+\alpha+2)-(-\alpha^2-\alpha) = -\alpha^2+2\alpha+2 \\
\alpha^{10} &= \alpha^9-\alpha^7 = (-\alpha^2+2\alpha+2)-(-2\alpha^2+1) = \alpha^2+2\alpha+1 \\
\alpha^{11} &= \alpha^{10}-\alpha^8 = (\alpha^2+2\alpha+1)-(2\alpha^2+\alpha+2) = 3\alpha^2+\alpha-1
\end{aligned}
$$

Now look back to the terms in (1) and the coefficients of α^2, α, and constant terms at the right in the above table of powers of α. Taking 0 to be t_0 as the term not listed before the three one's, we have the following pattern for the powers of alpha.

$$
\begin{aligned}
\alpha^3 &= t_2\alpha^2-t_0\alpha-t_1 \\
\alpha^4 &= t_3\alpha^2-t_1\alpha-t_2 \\
\alpha^5 &= t_4\alpha^2-t_2\alpha-t_3 \\
\alpha^6 &= t_5\alpha^2-t_3\alpha-t_4 \\
\alpha^7 &= t_6\alpha^2-t_4\alpha-t_5 \\
\alpha^8 &= t_7\alpha^2-t_5\alpha-t_6 \\
&\cdots
\end{aligned}
$$

Let the three roots of the characteristic equation, $x^3 - x^2 + 1 = 0$ be α, β, and γ. The above table of equations can be generalized as

$$
\begin{aligned}
\alpha^n &= t_{n-1}\alpha^2-t_{n-3}\alpha-t_{n-2} \\
\beta^n &= t_{n-1}\beta^2-t_{n-3}\beta-t_{n-2} \\
\gamma^n &= t_{n-1}\gamma^2-t_{n-3}\gamma-t_{n-2}
\end{aligned}
$$

Subtracting the second equation from the first and the third from the second, we have

$$a^n - \beta^n = t_{n-1}(\alpha^2 - \beta^2) - t_{n-3}(\alpha - \beta)$$
$$\beta^n - \gamma^n = t_{n-1}(\beta^2 - \gamma^2) - t_{n-3}(\beta - \gamma)$$

$$\frac{\alpha^n - \beta^n}{\alpha - \beta} = t_{n-1}(\alpha + \beta) - t_{n-3}$$

$$\frac{\beta^n - \gamma^n}{\beta - \gamma} = t_{n-1}(\beta + \gamma) - t_{n-3}$$

Subtracting the last two equations, solving for t_{n-1}, and replacing $(n-1)$ with n, we get

$$t_n = \frac{1}{\alpha - \gamma}\left(\frac{\alpha^{n+1} - \beta^{n+1}}{\alpha - \beta} - \frac{\beta^{n+1} - \gamma^{n+1}}{\beta - \gamma}\right) \tag{4}$$

In order to be able to graph (4), we need representations for α, β, and γ. The cubic formula which solves $x^3 + ax^2 + c = 0$ (with b, the coefficient of x, taken as 0) is

$$r_1 = \frac{-a}{3} - \frac{\sqrt[3]{2}(-a^2)}{3\left(-2a^3 + \sqrt{4(-a^2)^3 + (-2a^3 - 27c)^2} - 27c\right)^{1/3}} + \left(\frac{-2a^3 + \sqrt{4(-a^2)^3 + (-2a^3 - 27c)^2} - 27c}{54}\right)^{1/3}$$

$$r_2 = \frac{-a}{3} - \frac{(1 + i\sqrt{3})(-a^2)}{108\left(-2a^3 + \sqrt{4(-a^2)^3 + (-2a^3 - 27c)^2} - 27c\right)^{1/3}}$$
$$- \left(\frac{1 - i\sqrt{3}}{2}\right)\left(\frac{-2a^3 + \sqrt{4(-a^2)^3 + (-2a^3 - 27c)^2} - 27c}{54}\right)^{1/3}$$

$$r_3 = \frac{-a}{3} - \frac{(1 - i\sqrt{3})(-a^2)}{108\left(-2a^3 + \sqrt{4(-a^2)^3 + (-2a^3 - 27c)^2} - 27c\right)^{1/3}}$$
$$- \left(\frac{1 + i\sqrt{3}}{2}\right)\left(\frac{-2a^3 + \sqrt{4(-a^2)^3 + (-2a^3 - 27c)^2} - 27c}{54}\right)^{1/3}$$

We now substitute $a = -1$ and $c = 1$. Define $w = \sqrt[3]{\dfrac{\sqrt{621} + 25}{2}}$. Then the cubic formula yields

$$\alpha = \frac{1}{3}(1 - w - w^{-1})$$
$$\beta = \frac{1}{3}\left(1 + \left(\frac{1+i\sqrt{3}}{2}\right)w + \left(\frac{1-i\sqrt{3}}{2}\right)w^{-1}\right) \quad (5)$$
$$\gamma = \frac{1}{3}\left(1 + \left(\frac{1-i\sqrt{3}}{2}\right)w + \left(\frac{1+i\sqrt{3}}{2}\right)w^{-1}\right) = \overline{\beta}$$

after considerable simplification. Figure 1 was created using (4) with the values of (5) inserted.

We now present a graphical analysis of our n^{th} term formula which contains some surprises when n is allowed to vary over the set of real numbers. First we show (Figure 2) a graph of the imaginary component (coefficient of i) for $-20 \leq n \leq 6$ with n incremented by 0.01.

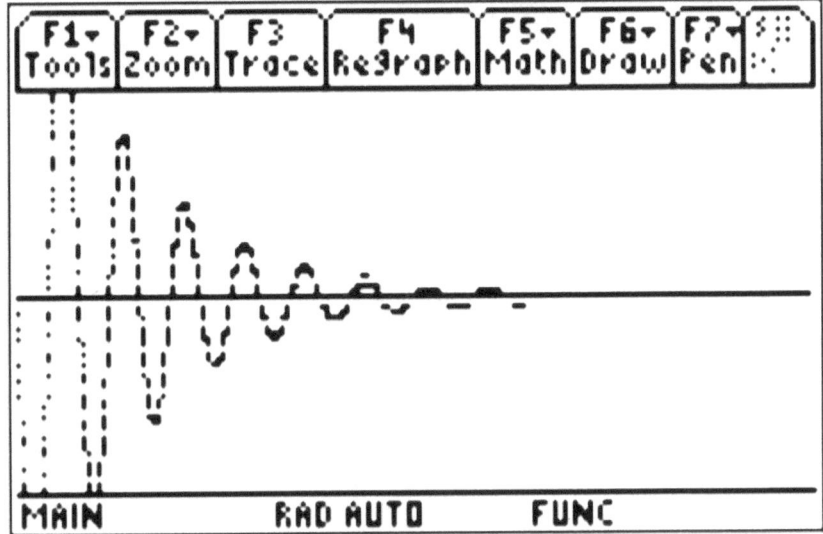

Figure 2

It is significant to note that what appears to be the x-axis is part of the graph of the t_n formula. The graph was drawn in the axes off mode. This means that $t_n = 0$ for many values of $n \leq 0$, and for *all* $n \geq 0$, $t_n = 0$.

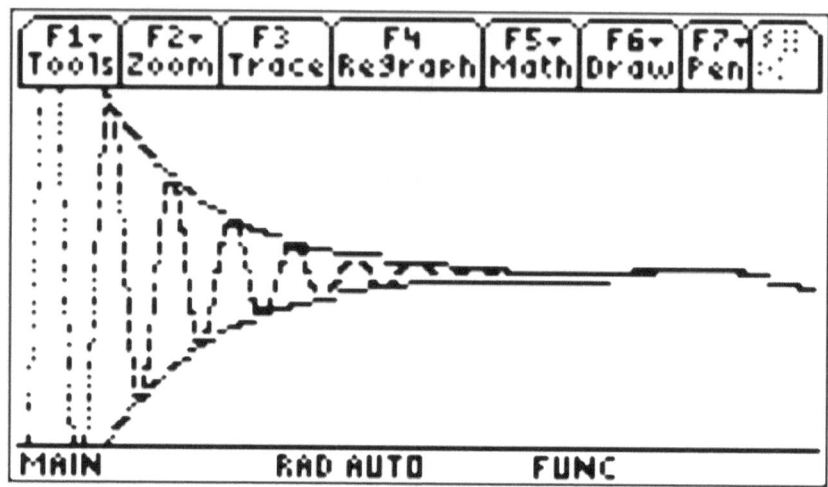

Figure 3

The next graph, Figure 3, is surprising. It indicates the graph of the real component of t_n for $-20 \leq n \leq 6$. In Figure 4 the same graph is shown except that $-20 \leq n \leq 20$. Interestingly, the envelope containing the oscillating graph for $n \leq 0$ is also part of the graph! As can be seen in Figure 4 especially, the graph follows the pattern of the sequence dots shown in the right half of Figure 1. The meeting points between the oscillating portion and the envelope coincide with the sequence dots in the left half of Figure 1. While watching the graphs of Figures 3 and 4 being drawn, the left hand portions appear as if three functions are being drawn simultaneously which then merge when n exceeds zero.

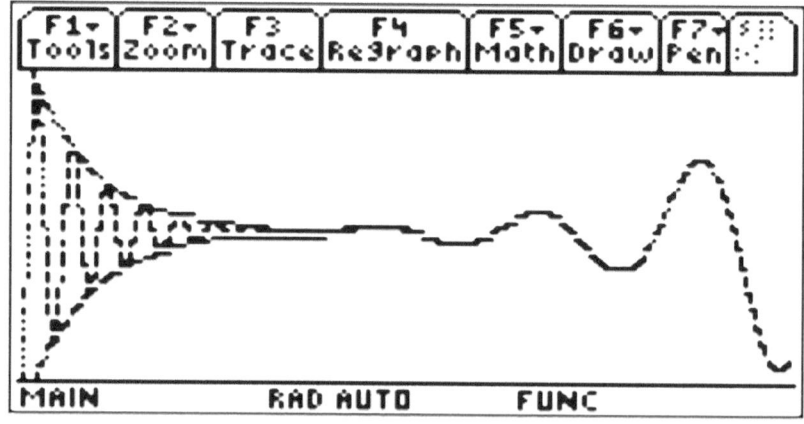

Figure 4

Figures 2 through 4 were drawn for all values of t_n with the only filter being the imaginary component for Figure 2 and the real component for Figures 3 and 4. Next we impose a filter in which the real component of t_n is drawn *only* if the imaginary component is zero. The result is shown in Figure 5 which should be compared with Figure 3 since the domain is $-20 \leq n \leq 6$. While the envelope is still included, only the rising component of the oscillating feature is seen. Figure 6 was drawn with the opposite filter to that employed in Figure 5. That is, the imaginary component of t_n is drawn *only* if the real component is zero.

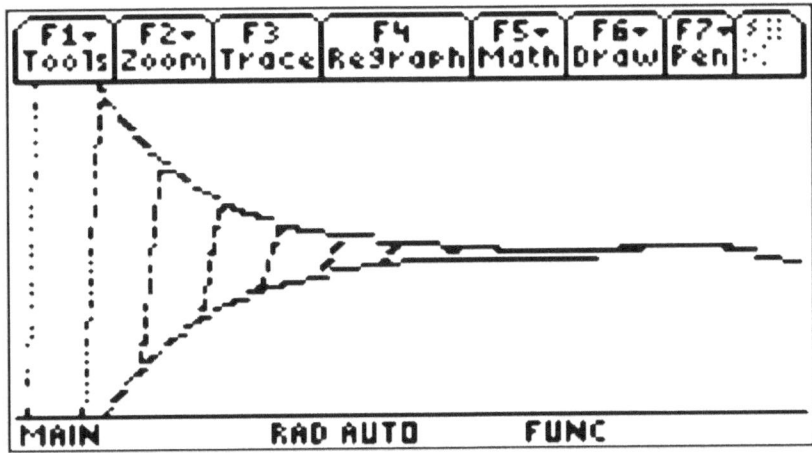

Figure 5

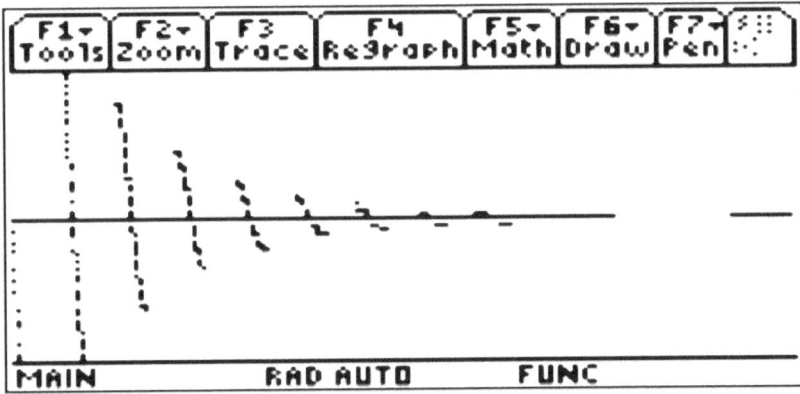

Figure 6

The envelope is missing but many points are plotted on the horizontal axis. The horizontal void exists near the right end since there is a portion of the Figure 5 graph in this location. The falling portion of the oscillating feature in the left hand side of Figure 2 also appears.

Another View

Thinking about all of the previous *Figures*, there is an alternate way to visualize what is the true nature of the t_n formula for all real values of *n*. Before describing this further, look at Figure 7, then put this article aside and see if you understand what is happening.

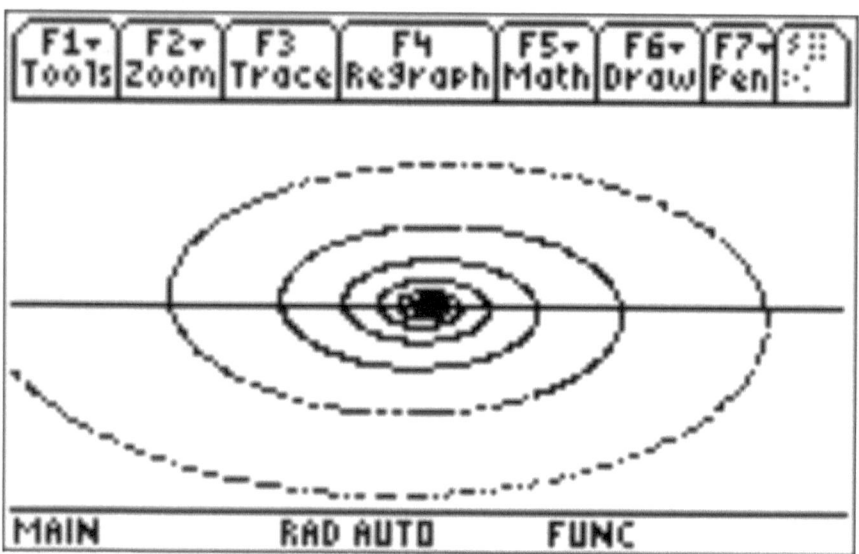

Figure 7

This is a very different view of the t_n function – it is in three dimensions. As you now know, t_n is complex-valued for $n < 0$ and real-valued for $n \geq 0$. Thus we have $t_n = Z(n) = x + yi$. In Figure 7 you have your eye in the n-axis. That is, the n-axis is perpendicular to the plane of the page containing Figure 7. Figure 7 itself is in three dimensions. The real *x*-axis is horizontal through the middle of the Figure with the positive *x*-axis toward the right. The horizontal line is actually the envelope appearing in Figures 3 and 4 seen edge-on. The imaginary *y*-axis is vertical, positive axis upward. The spiral begins way behind the page (20 units) and has its tip end touching the page at the origin $(n, x, y) = (0, 0, 0)$. Looking at the spiral, it is as if we were at the base of a tornado looking up from the ground.

For large negative n, the graph is a spiral of large diameter tangent to the curving envelope which lies in the *nx*-plane. As *n* increases, the spiral decreases in radius, with the radius becoming near zero at the origin. Thus the spiral and both branches of the envelope tend to converge as the graph nears and passes the origin. For increasing positive *n*, the graph oscillates right and left with increasing amplitude in the *nx*-plane. From our viewpoint in Figure 7, this oscillating curve lies in the plane of the horizontal line. Figure 8 shows how Figures 7 and 4 can be put together to illustrate the perspective described above.

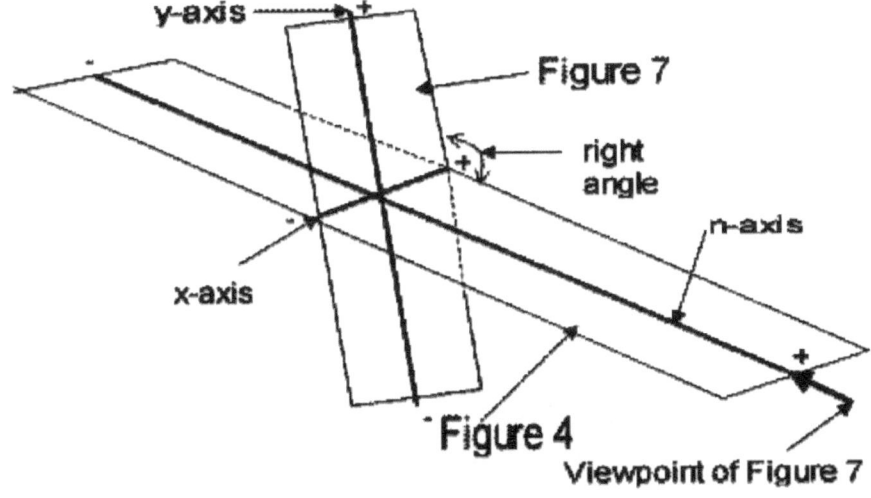

Figure 8

To provide a three dimensional graph of $t_n = Z(n) = x + yi$, a special program for the TI 89 had to be devised mapping (n, x, y) onto (X,Y). Figure 9 shows how this was done. The screen origin, $(0,0) = (X,Y)$, was placed at the center of the screen for simplicity, and corresponds to $(0,0,0) = (n, x, y)$. Thus the geometry of Figure 9 causes angle AOB to be $45°$ and the distances from O to B to be equal to that from C to D. We need to compute the values of X and Y from the values of n, x, and y using a little trigonometry. The value of X is the perpendicular distance from O to line segment FE (not shown) which is equal to DE, and $Y = FE - OD = FE - (OC + CD)$.

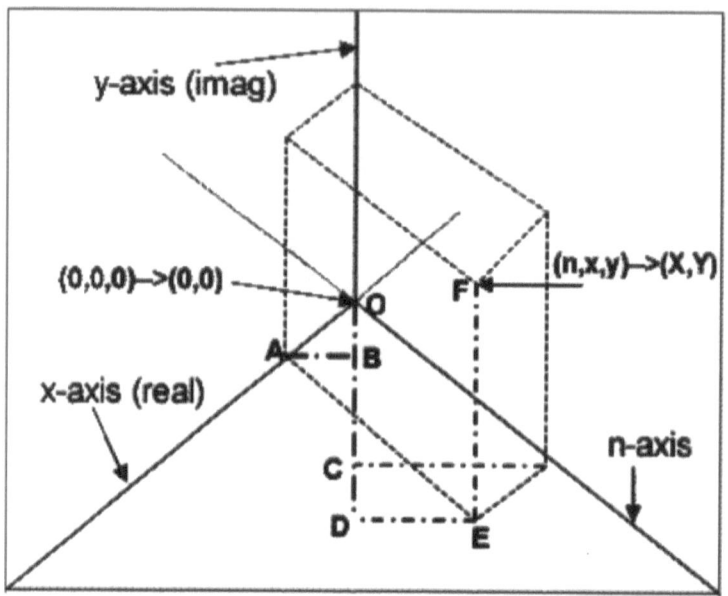

Figure 9

Therefore we have

$$Y = y - (x \cos 45° + n \cos 45°)$$
$$X = x \sin 45° + n \sin 45°$$

$$Y = y - \frac{n+x}{\sqrt{2}}$$
$$X = \frac{n-x}{\sqrt{2}}$$

In the actual program for the TI 89, x was replaced with $-x$ to create the diagram shown in Figure 10 since the positive x-axis, as indicated in Figure 8, is into the page. The viewpoint of Figure 10 is the same as that in Figure 8. The lighter density of pixels indicates greater distance behind the page.

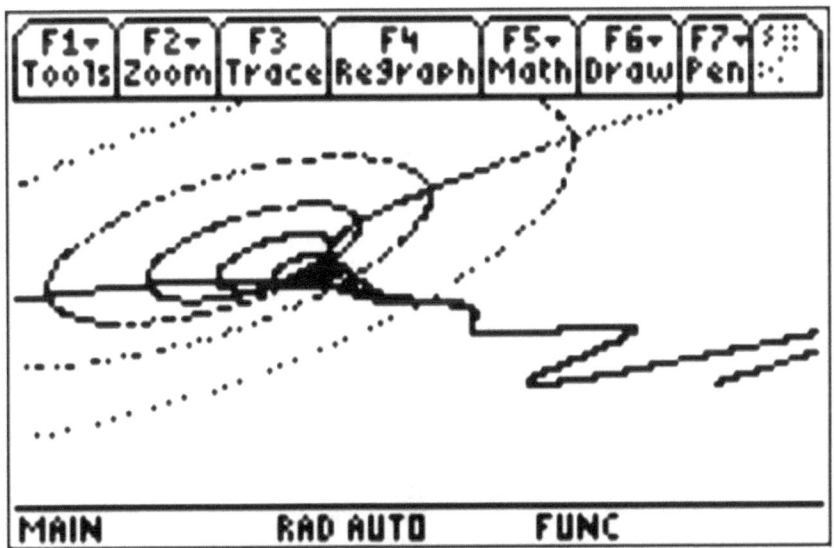

Figure 10

Conclusion

The author discovered the sequence in the early 1970's. At that time there was no readily available computing power to draw the graphs in the Figures. Further, the prospect of having to solve a cubic equation seemed too daunting at the time, so no practical solution was thought possible. As a secondary school teacher, the author has enjoyed presenting the sequence to students learning signed numbers as an application. Once some of the terms are generated, they provide good motivation for a graphical representation such as that shown in Figure 1.

After an early attempt failed to produce a simple formula for t_n, which we now know does not exist, the author tried to gain insight by examining a much simpler sequence

$$1,1,0,-1,-1,0,1,1,0,\ldots$$

The development of its t_n formula resulted in the creation of a meteorological application in describing climate with respect to daily average temperatures. (See "An Analysis of a Simple Periodic Sequence", *THE MATHEMATICS TEACHER*, March, 1972, page 255.) This book contains an article describing how the simpler sequence relates to average temperatures.

Aircraft Surveillance of Speeders – Realistic Case

Introduction

In many editions of the Thomas Calculus textbook (for example, p. 221, problem #34 in *Thomas' Calculus, 11th Edition,* 2005, Pearson Education, Inc.) a related rates problem appears which was taken from an article written by the author in *The Mathematics Teacher*. The textbook version of the problem is the simplified version with the plane in level flight and the car on a level highway. Incidentally, in the original version of the simplified problem, the speed of the car was 62 mph which was excessive since the problem was conceived in the 1970's at a time when the national speed limit was 55 mph, presumably to save gasoline! Here we present the more realistic case with the plane on a descending line path and the car moving downhill.

The diagram below shows the scenario whereby an aircraft in non-level flight is observing a car moving along a non-level highway. Doppler radar on board the aircraft can indicate both the line of sight distance to a target and the rate of change in this distance. It can also determine the distance straight downward to the road and the rate of change in this distance. Circuitry aboard the aircraft can also determine the angle, θ, and its rate of change at the instant the distance observations are made. The aircraft is flying at the angle of depression of α while the road is inclined at an angle of elevation of Φ.

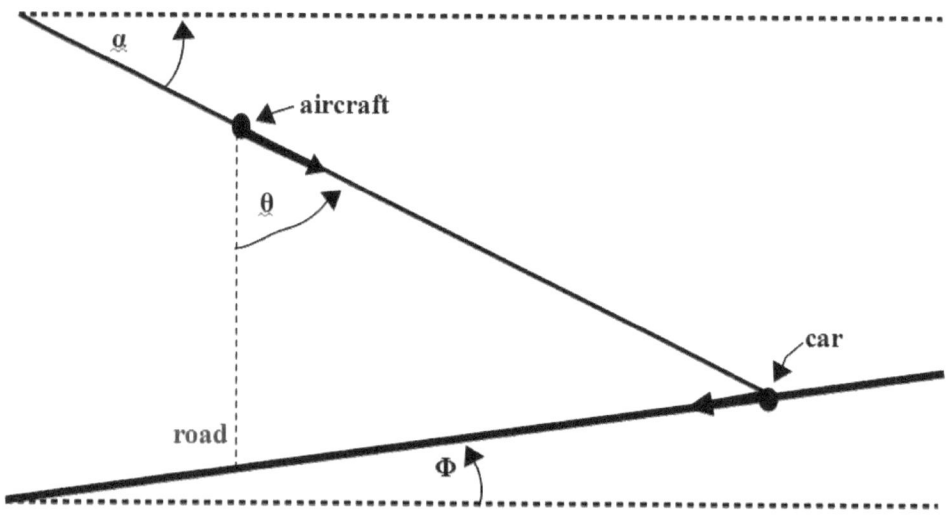

Figure 1

Figure 1 depicts the given information and configuration of the problem. Subsequently Figure 2 provides additional details involving defined variables.

The Problem

A plane, flying along a path at the angle of depression of α spots a car with Doppler radar and determines that at the instant the distance is S, the rate of change in S is $\frac{dS}{dt}$. The aircraft also directs the Doppler radar straight downward and determines that the elevation is h and the rate of change in the elevation is $\frac{dh}{dt}$. The problem is to determine the rate of the car, $\frac{dx}{dt}$.

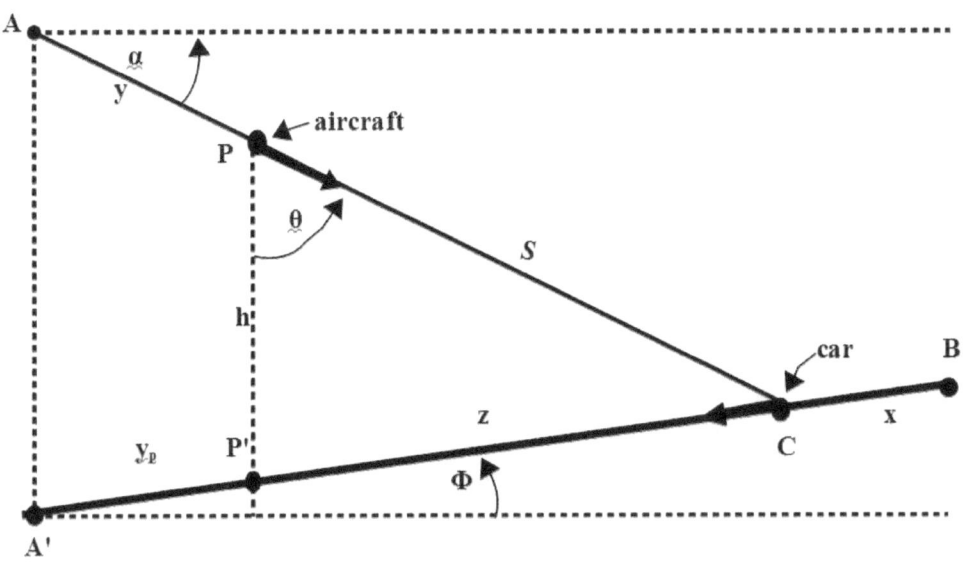

Figure 2

The Solution

At some arbitrary time before the aircraft made its observations place the aircraft at point A and the car at point **B**. At the time of observation place the aircraft at point **P** which is at the distance **y** from point **A**, and place the car at point **C** which is at the distance of **x** from point **B**. Let points **A'** and **P'** be the vertical projections of points **A** and **P**, respectively, onto the road. Let y_p be the distance from Point **A'** to point **P'**. Let **z** be the distance from **P'** to **C** and S be the distance from **P** to **C** along the line of sight. Let θ be the measure of **angle CPP'**. The elevation of the aircraft at the instant of observation is **PP'** = **h**, and the aircraft's speed is $\frac{dy}{dt}$. Thus the eight known values at the instant the observation is done are **h, S, θ, α,** $\frac{dy}{dt}, \frac{dh}{dt}, \frac{dS}{dt}$, and $\frac{d\theta}{dt}$.

Clearly as time elapses the distance from **A'** to **B** is constant. Let **A'B = K** where **K** is a constant. Then it follows that

$$y_p + z + x = K$$

Differentiating both sides with respect to time, t, we find

$$\frac{dx}{dt} = -\frac{dy_p}{dt} - \frac{dz}{dt}. \tag{1}$$

The solution can be completed when we determine $\frac{dy_p}{dt}$ and $\frac{dz}{dt}$ in terms of any of the eight known values, **h, S, θ, α,** $\frac{dy}{dt}$, $\frac{dh}{dt}$, $\frac{dS}{dt}$, and $\frac{d\theta}{dt}$. Let's attack $\frac{dy_p}{dt}$ first which requires determining the relationship between **y** and **yp**. In order to do this we need to focus on the left part of Figure (2) which is shown in Figure 3.

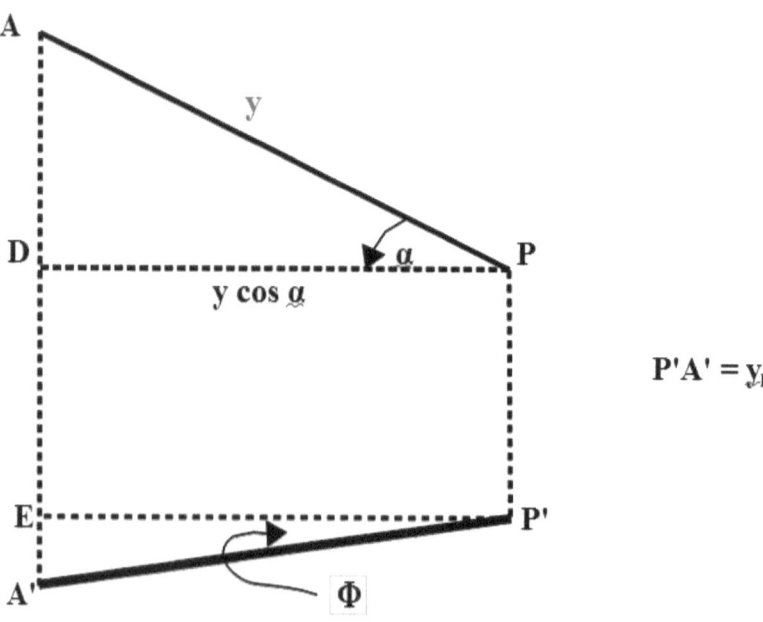

Figure 3

In triangle **APD** we see that $DP = y \cos \alpha$, and from triangle **EP'A'** we have

$$y_p = (EP') \sec \Phi = (DP) \sec \Phi = y(\cos \alpha)(\sec \Phi).$$

Therefore we have an expression for $\dfrac{dy_p}{dt}$ by differentiating the above y_p statement.

$$\frac{dy_p}{dt} = (\cos \alpha)(\sec \Phi) \frac{dy}{dt} \tag{2}$$

We next focus on the right side of the diagram which is shown on the next page in Figure 3. We want to eliminate Φ from equation (2) since it cannot be easily measured from the aircraft. With $\overline{CQ} \perp \overline{PP'}$ we can use right triangle trigonometry. Since angle **P'CQ** is Φ, from right triangle **P'CQ** we have $CQ = z \cos \Phi$. From right triangle **PQC** we know that $\sin \theta = \dfrac{z \cos \Phi}{S}$ so that $\sec \Phi = \dfrac{z}{S \sin \theta}$. Substituting this into (2) results in equation (3).

$$\frac{dy_p}{dt} = \frac{z \cos \alpha}{S \sin \theta} \frac{dy}{dt} \tag{3}$$

Next we turn our attention to the second derivative in the right side of (1), $\dfrac{dz}{dt}$. Look back to Figure 1 and triangle **PP'C**. By the Law of Cosines,

$$z^2 = h^2 + S^2 - 2hS(\cos \theta)$$

Differentiating both sides with respect to **t** we have

$$\frac{dz}{dt} = \frac{1}{z} \left[h \frac{dh}{dt} + S \frac{dS}{dt} - S(\cos \theta) \frac{dh}{dt} - h(\cos \theta) \frac{dS}{dt} + hS(\sin \theta) \frac{d\theta}{dt} \right] \tag{4}$$

after dividing both sides by 2**z**. We substitute the expressions for the derivatives in (4) and (3) into equation (1) to obtain $\dfrac{dx}{dt}$, the speed of the car, in terms of the eight quantities observed by the aircraft. (*NOTE:* The following expression is divided into three segments for proper Kindle viewing in portrait mode.)

$$\frac{dx}{dt} = \frac{1}{z} \left[\frac{dh}{dt}(S(\cos \theta) - h) + \frac{dS}{dt}(h(\cos \theta) - S) - \frac{d\theta}{dt}(hS)(\sin \theta) \right] - \frac{z \cos \alpha}{S \sin \theta} \frac{dy}{dt}$$

where

$$z = \sqrt{h^2 + S^2 - 2hS(\cos\theta)}$$

Note that a TI84 programmable graphing calculator on board the aircraft receiving input of the eight observed values can provide the car's speed in about a second – plenty of time to radio a hiding patrol car on the road to make an arrest, should one be necessary.

An Artillery Challenge

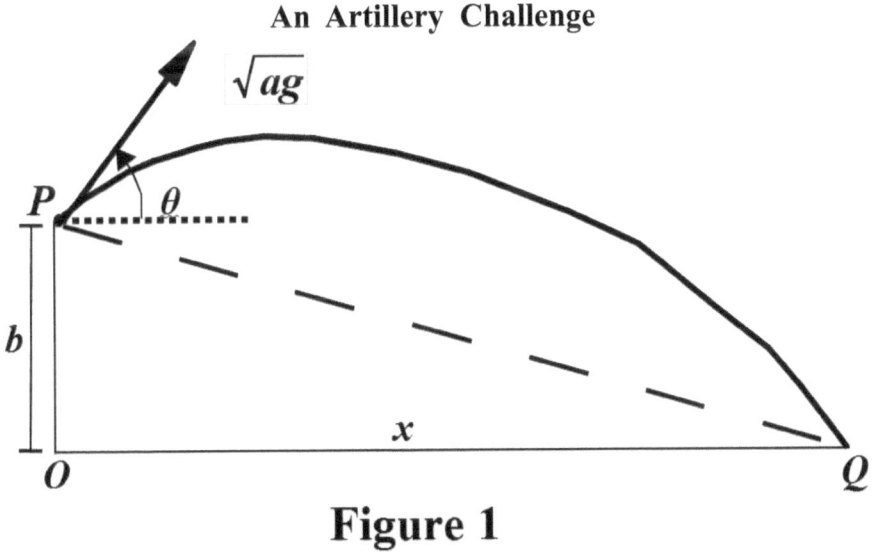

Figure 1

(a) A soldier on a hill at position P fires a mortar shell with speed $\sqrt{ag}$ at an angle θ above the horizontal, where a is a constant. P is at a vertical height b above the surrounding land which is a horizontal plane. The shell strikes the plane at the point Q, and O is the point at the level of the plane vertically below P, as shown in the diagram. Letting $OQ = x$, obtain the equation

$$x^2 \tan^2 \theta - 2ax \tan\theta + (x^2 - 2ab) = 0.$$

(b) Show that the maximum value of x, as θ varies, is $\sqrt{a(a+2b)}$ and that this is achieved when

$$\tan\theta = \sqrt{\frac{a}{a+2b}}.$$

(c)   The sound of the shell being fired travels along the straight line PQ at a constant speed $\sqrt{cg}$. Given that the shell is fired to achieve its maximum range, show that if a person standing at Q hears the sound of firing before the shell arrives at Q, so giving him/her time to take cover, then

$$c > \tfrac{1}{2}(a+b).$$

Solution:

(a) First we derive the parametric equations for the path of the shell. We have the initial conditions:

$t = 0$ which implies all the following:
$$x = 0, y = 0,$$
$$v_x = \sqrt{a\,g}(\cos\theta), \quad v_y = \sqrt{a\,g}(\sin\theta)$$
$$a_x = 0, \; a_y = -g$$

Proceeding with the integrations beginning with the accelerations we find:

$$v_y = -gt + \sqrt{a\,g}(\sin\theta), \quad v_x = \sqrt{a\,g}(\cos\theta)$$

$$y = -\frac{1}{2}gt^2 + t\sqrt{a\,g}(\sin\theta) + b, \quad x = t\sqrt{a\,g}(\cos\theta)$$

Solving the x expression for t and substituting into the y function we have

$$y = -\frac{1}{2}g\left(\frac{x^2}{a\,g\cos^2\theta}\right) + x\tan\theta + b$$

When $OQ = x$ the value of $y = 0$. Substituting this into the equation of the path we find

$$\frac{-x^2}{2a\cos^2\theta} + x\tan\theta + b = 0$$

$$\frac{-x^2}{2a}\sec^2\theta + x\tan\theta + b = 0$$

$$-x^2(1 + \tan^2\theta) + 2ax\tan\theta + 2ab = 0$$

$$x^2\tan^2\theta - 2ax\tan\theta + \left(x^2 - 2ab\right) = 0 \qquad (1)$$

(b) We differentiate implicitly both sides of this equation with respect to θ and then set $\dfrac{dx}{d\theta} = 0$.

$$D_\theta\left(x^2\tan^2\theta - 2ax\tan\theta\right) + D_\theta\left(\left(x^2 - 2ab\right)\right) = 0$$

$$D_\theta\left(x^2\sec^2\theta - 2ax\tan\theta - 2ab\right) = 0$$

$$2x\left(\frac{dx}{d\theta}\right)\sec^2\theta + 2x^2\sec^2\theta\tan\theta - 2a\tan\theta\left(\frac{dx}{d\theta}\right) - 2ax\sec^2\theta = 0$$

Setting $\dfrac{dx}{d\theta} = 0$,

$$2x^2\sec^2\theta \tan\theta - 2ax\sec^2\theta = 0$$

$$(x\sec^2\theta)(x\tan\theta - a) = 0$$

Since neither x nor $\sec^2\theta$ can equal zero we have the relation

$$x = \frac{a}{\tan\theta}.$$

Substituting this into equation (1) yields

$$a^2 - 2a\left(\frac{a}{\tan\theta}\right)\tan\theta + \frac{a^2}{\tan^2\theta} - 2ab = 0$$

$$\frac{a^2}{\tan^2\theta} + \left(-a^2 - 2ab\right) = 0$$

$$\tan^2\theta = \frac{a^2}{a^2 + 2ab}$$

$$\tan\theta = \sqrt{\frac{a}{a+2b}}.$$

Returning to the relation $x = a/(\tan\theta)$ we see that

$$x = a\sqrt{\frac{a+2b}{a}}$$

$$x_{MAX} = \sqrt{a(a+2b)}$$

c) Clearly point P is at $(0, b)$ and point Q is at $\left(\sqrt{a(a+2b)}, 0\right)$. Thus the distance from P to Q is

$$PQ = \sqrt{a^2 + 2ab + b^2} = a + b.$$

Therefore the time it takes to hear the shell being fired is given by $\frac{a+b}{\sqrt{cg}}$. To obtain the time it takes for the shell to reach point Q we go back to the equation for x, namely $x = t\sqrt{ag}(\cos\theta)$, and replace x with x_{MAX} to obtain

$$\sqrt{a(a+2ab)} = t\sqrt{ag}(\cos\theta)$$

$$t = \frac{\sqrt{a(a+2b)}}{\sqrt{ag}\cos\theta} \qquad (2)$$

We employ the expression $\tan\theta = \sqrt{\dfrac{a}{a+2b}}$ obtained earlier to get the value of $\cos\theta$ to insert into the time equation (2). Consider the reference triangle in Figure 2 based on this tangent value for θ.

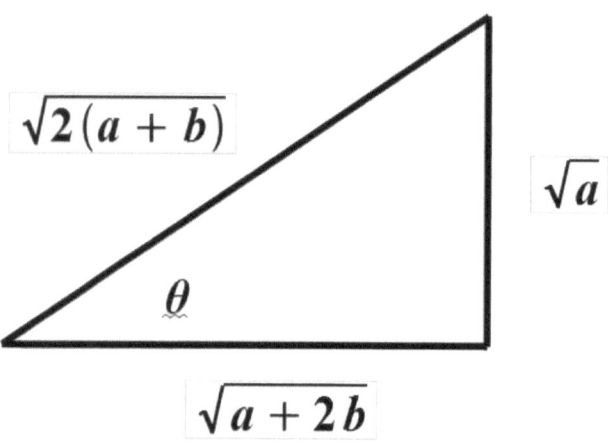

Figure 2

$$t = \dfrac{\sqrt{a(a+2b)}}{\sqrt{ag}\cos\theta} = \dfrac{\sqrt{a(a+2b)}}{\sqrt{ag}\left(\dfrac{\sqrt{a+2b}}{\sqrt{2(a+b)}}\right)} = \dfrac{\sqrt{2a(a+b)(a+2b)}}{\sqrt{ag}(a+2b)}$$

$$t = \sqrt{\dfrac{2(a+b)}{g}}$$

The person at point Q needs to hear the shell being fired *before* the shell arrives at Q. Thus we have the inequalities which finally simplify to (3).

$$\dfrac{a+b}{\sqrt{cg}} < \sqrt{\dfrac{2(a+b)}{g}}$$

$$\dfrac{(a+b)^2}{cg} < \dfrac{2(a+b)}{g}$$

$$\dfrac{a+b}{2} < c \quad (3)$$

Proving the Focusing Properties of the Parabola and Ellipse Without Calculus!

Introduction

The conic sections which tend to most fascinate students are the parabola and ellipse. Their focusing properties are interesting and can be demonstrated with models. For the parabola there are magnifying mirrors, radar and satellite antennas, directional microphones, and reflecting telescopes. For the ellipse there are various whispering chambers such as the Rotunda Building in Washington D.C. and the elliptical dome of Saint Paul's Cathedral in London, England. Ripple tanks can be designed to show the focusing properties, too. Even an elliptical Chinet dish filled with some water is effective. If a finger is dipped at one focus, especially if the lighting is slanted, it will cause waves to converge at the other focus.

Proving the focusing properties is another matter. Proofs are not readily available as many textbooks presenting conic sections to the pre-calculus student state that proofs are beyond the scope of the course as calculus is needed. Indeed, the most popular proofs do employ calculus. In fact, some calculus texts will ask students to supply focus proofs as exercises.

The author recalls in his own college calculus course as a student, more years ago than he cares to remember, when he asked the professor if there were a non-calculus means of establishing that a parabolic reflector will converge rays of light which are parallel to the axis of symmetry to the focus point. He was told that it was probably not possible. This briefly bothered him at the time, but his unease was forgotten for years until he became a high school teacher. Anticipating this question from one of his students, proofs were discovered using elementary plane geometry for the parabola and ellipse focus properties.

Beyond the basic geometry, the reader only needs to know two principles of optics. The first is that the angle of incidence ($\angle 1$) equals the angle of refraction ($\angle 2$). See Figure 1. This also means, of course, that $\angle 3 = \angle 4$ as complements of congruent angles are congruent. The second principle of

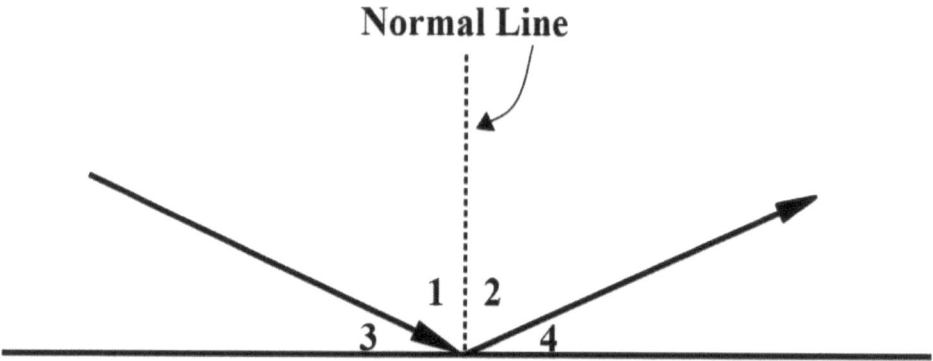

Figure 1

optics concerns the direction of reflection caused by curved mirrors. At the point of impact light behaves as if the mirror were flat along the line tangent to the curve at the impact point as shown in Figure 2. Note that ∡1=∡2 since the incoming ray and the outgoing ray make equal angles with the tangent line, L, which is tangent to the curved mirror at impact point P.

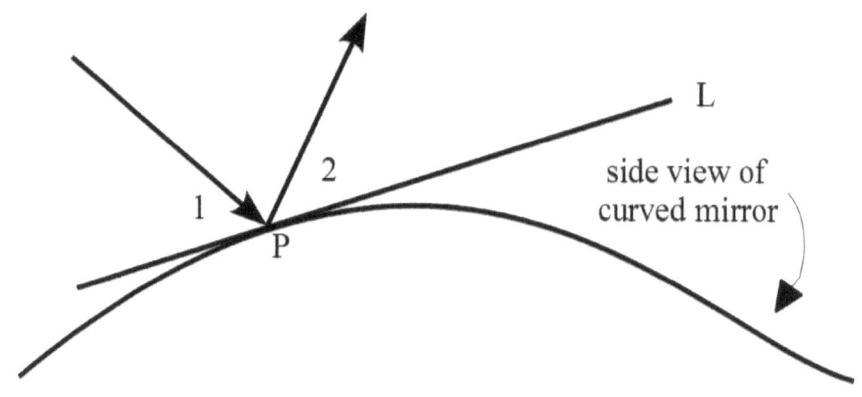

Figure 2

Proof of the Parabola Focusing Property

In Figure 3 a parabola is drawn having the directrix, L, and focus, F. The incoming ray VP would intersect L at right angles at point R if extended, but it strikes the parabola at point P. Assume that the focusing property does not hold. That is, if a tangent line is drawn tangent to the parabola at point P, the ray reflecting from point P will not pass through point F. We now draw the line through P which does reflect the ray through point F. Since the tangent line at P is unique, the focusing line must

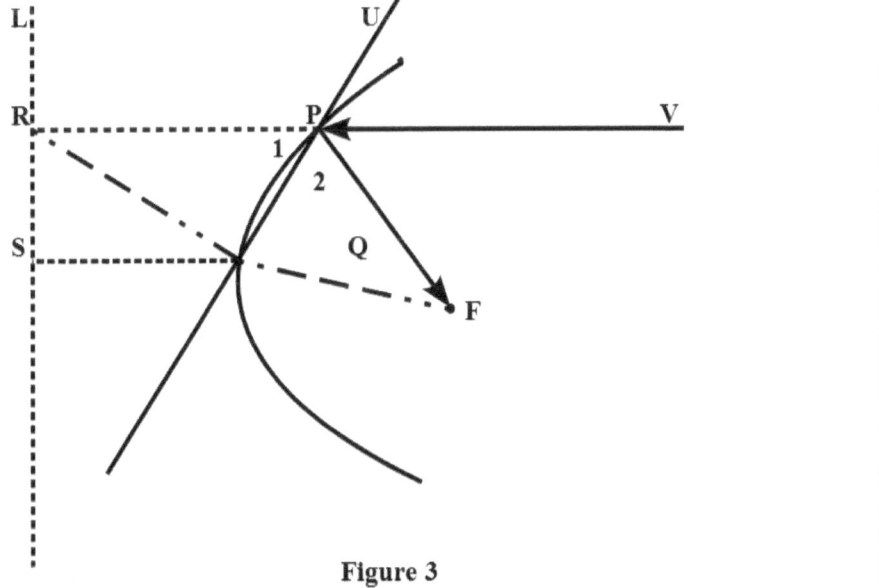

Figure 3

intersect the parabola at a second point, call it Q. Draw $\overline{QR}$ and $\overline{QF}$. $\sphericalangle UPV = \sphericalangle 2$ by the reflection property, and since $\sphericalangle UPV$ and $\sphericalangle 1$ are vertical we have $\sphericalangle 1 = \sphericalangle 2$. By the definition of parabola, PF = PR, and since PQ = PQ, triangle PQF is congruent to triangle PQR by SAS. The corresponding sides QR = QF. Draw $\overline{QS} \perp L$. Applying the definition of parabola to point Q we see that QF = QS so that QR = QS making triangle QRS isosceles. Thus this triangle has two right angles, one at point S and the other at point R which is a contradiction! This establishes that only the tangent line can reflect a ray which is perpendicular to the directrix (if extended) through the focus.

The Ellipse Focusing Property

Suppose a ray coming from point F_2 strikes an elliptic reflector at any point P on the ellipse as shown in Figure 4. The ray will reflect from P as if the reflector were straight rather than curved at P. If a tangent line were drawn at P, the incoming and reflected ray would make equal angles with the tangent line.

To establish the focus property we must prove that if any line segment is drawn from F_2 to P and a second line segment is drawn from P to F_1, these two line segments must make equal angles with the tangent line drawn to the ellipse at point P. We use an indirect proof as follows.

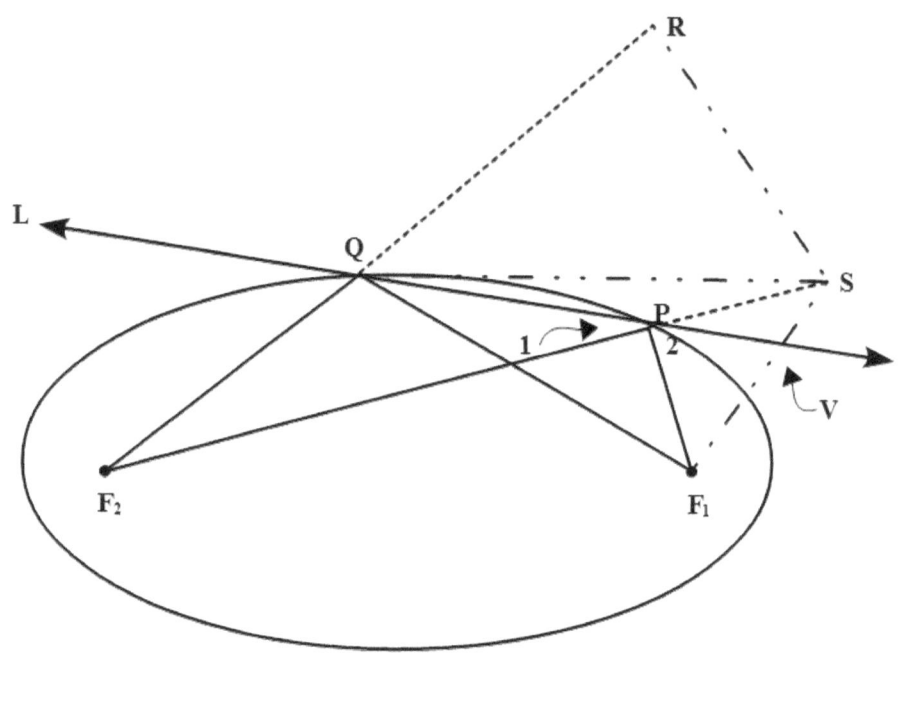

Figure 4

Suppose equal angles are NOT made by $\overline{F_2P}$ and $\overline{PF_1}$ with the tangent line at P. Draw line L through P so that equal angles are made with these two segments. Thus ∡1=∡2. L is not tangent by supposition, so let point Q be the second point of intersection of line L with the ellipse. Now draw $\overline{F_2Q}$ and $\overline{QF_1}$. Extend $\overline{F_2Q}$ to point R so that $QF_1 = QR$, and extend $\overline{F_2P}$ to point S so that $PF_1 = PS$. By the definition of an ellipse, $F_2P + PF_1 = F_2Q + QF_1$ so that by substitution, $F_2R = F_2S$ making triangle F_2RS isosceles. Draw $\overline{SF_1}$. Triangle PSF_1 is also isosceles. Let the intersection of line L with $\overline{SF_1}$ be point V. Since ∡1 and ∡SPV are vertical, they are congruent so that ∡2=∡SPV. Therefore line L is the perpendicular bisector of $\overline{SF_1}$ so that every point of line L is equidistant from points S and F_1. Draw $\overline{QS}$. Point Q is equidistant from S and F_1 so that $QF_1 = QS$ making $QR = QS$. Thus triangle QRS is isosceles and ∡QRS=∡QSR. But earlier we saw that triangle F_2RS is also isosceles making ∡QRS=∡PSR. This implies that ∡PSR=∡QSR. This contradicts that points P and Q are distinct points thus establishing the focusing property of the ellipse.

A Curious Fountain Problem

A fountain is to consist of multiple nozzles at ground level at one location, each able to aim in any direction and to squirt a substantial jet of water at constant velocity, v_0. We will neglect air resistance. Using g as the gravitational constant, find an expression in terms of g and v_0 for the total volume of space within reach of the fountain. Clearly if you were designing a public fountain like this, you would want to fence the perimeter where this space intersects the ground so as to avoid wetting the public. Neglecting air resistance actually enhances the situation as we would want to slightly over-estimate the wettable volume in fencing off the fountain.

Solution:

The solution will be in two phases.

(1) We need to consider a plane perpendicular to the ground containing the nozzle point, **N**. Placing the origin at **N** in a coordinate system in this plane, we will determine an equation of the set of points at the critical distance from **N**. These points lie on a curve and are at the furthest possible distance which can be struck by the fountain. See Figure 1 for a representation of this enveloping curve together with some representative water jet trajectories.

(2) Rotating the area under this curve about the y axis creates the volume of the wettable region. Calculating the volume will solve the problem.

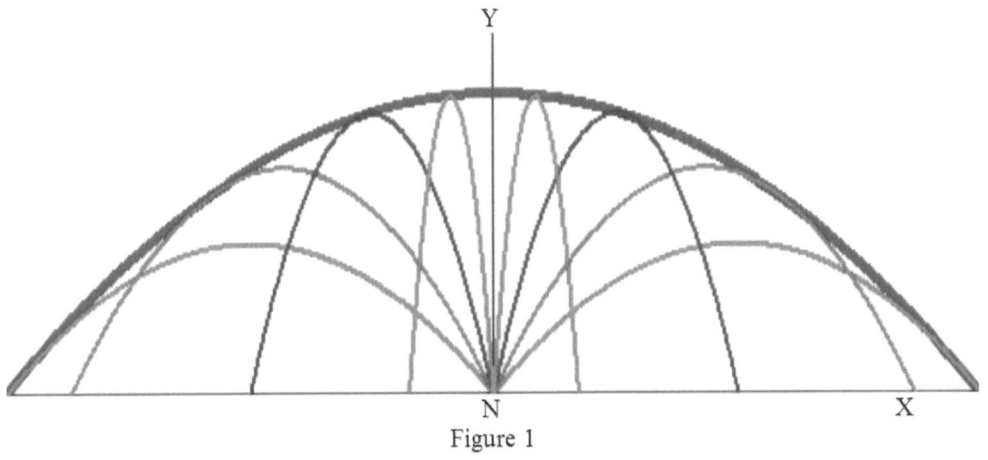

Figure 1

Figure 1 shows eight representative paths for water jets coming from point **N** which all remain within the blue curve. This thick curve, called an ***envelope curve***, is the curve of points whose equation we need to determine. Notice that the envelope curve does ***not*** contain the high points in each individual trajectory, an observation particularly obvious from an examination of the lower trajectories. All trajectories do appear, however, to be tangent to the envelope.

The key idea in deriving the equation of the envelope is the following characterization of the points in the upper half-plane of the diagram above the *x*-axis. Points above the envelope are on none of the fountain trajectories. Each point below the envelope lies on exactly two trajectories. *Each point on the envelope lies on **exactly one** trajectory.*

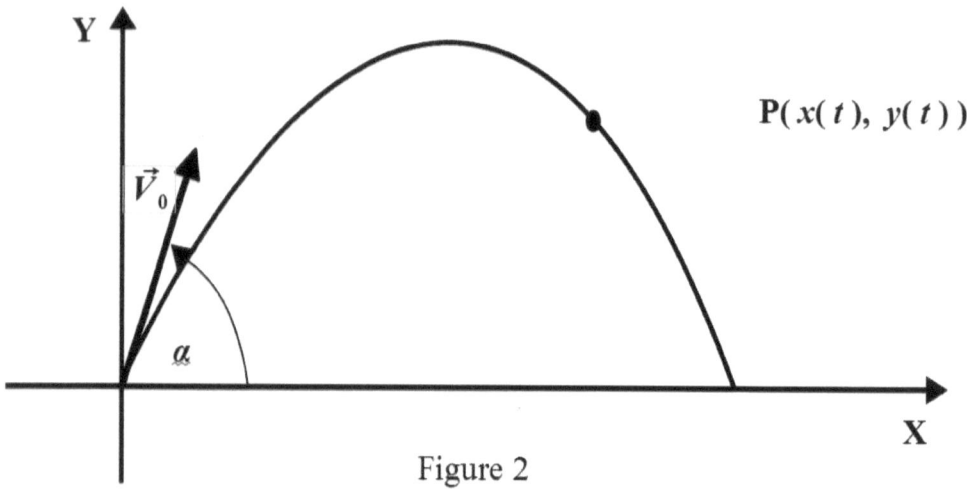

Figure 2

We now derive parametric equations for the trajectories. Figure 2 shows a representative trajectory for a water jet originating at $(x, y) = (0,0)$ at time $t = 0$, initial velocity in the *x*-direction of $V_x = \|V_0\|\cos\alpha$ and initial velocity in the *y*-direction of $V_y = \|V_0\|\sin\alpha$. The initial horizontal and vertical components of the acceleration are $a_x = 0$ and $a_y = -g$. We begin with two differential equations by using the definition of acceleration.

$$a_x = \frac{dv_x}{dt} = 0 \text{ and } a_y = \frac{dv_y}{dt} = -g$$

Solving these in parallel fashion,

$$v_x = \int 0\, dt = C_1 \text{ and } v_y = \int -g\, dt = -gt + C_2$$

$t = 0, v_x = v_0 \cos\alpha$ implies that $C_1 = v_0 \cos\alpha$

$t = 0, v_y = v_0 \sin\alpha$ implies that $C_2 = v_0 \sin\alpha$

Substituting the constants and then using the definition of velocity we have

$$v_x = \frac{dx}{dt} = v_0 \cos\alpha \text{ and } v_y = \frac{dy}{dt} = -gt + v_0 \sin\alpha$$

Separating variables and integrating,

$$x = v_0 \cos\alpha \int dt = v_0 t \cos\alpha + C_3$$

and

$$y = \int (-gt + v_0 \sin\alpha) dt = -\frac{1}{2} gt^2 + v_0 t \sin\alpha + C_4$$

Letting $t = 0$, $x = 0$, and $y = 0$ we obtain $C_3 = 0$ and $C_4 = 0$. Putting all of this together we have the following set of parametric equations describing the trajectory.

$$x(t) = v_0 t \cos\alpha \text{ and } y(t) = -\frac{1}{2} gt^2 + v_0 t \sin\alpha$$

Since we want to eliminate t, we use the top equation to get an expression for t which is then substituted into the second equation.

$$y = -\frac{1}{2} g \left(\frac{x}{v_0 \cos\alpha} \right)^2 + v_0 \left(\frac{x}{v_0 \cos\alpha} \right) \sin\alpha$$

$$y = \left(\frac{-gx^2}{2v_0^2} \right) \sec^2\alpha + x \tan\alpha$$

$$y = \left(\frac{-gx^2}{2v_0^2} \right)(1 + \tan^2\alpha) + x \tan\alpha \tag{1}$$

As α varies from 0 to π, all possible trajectories are generated by equation (1). Remember that for each point $P(x, y)$ inside the envelope, two trajectories pass through **P** while for each point *on* the envelope, **exactly one** point **P** lies on a trajectory. Stated in another way, to each α we want **exactly one** pair of x and y values to correspond. We observe that equation (1) is quadratic in the variable $\tan(\alpha)$ so we rewrite (1) in the form

$$A(\tan\alpha)^2 + B(\tan\alpha) + C = 0$$

and then solve for $(\tan\alpha)$ in terms of x and y.

$$2v_0^2 y = -gx^2 - gx^2 \tan^2\alpha + 2v_0^2 x * \tan\alpha$$

$$gx^2 \tan^2\alpha - 2v_0^2 x * \tan\alpha + (2v_0^2 y + gx^2) = 0$$

Using the quadratic formula we solve for $(\tan\alpha)$.

$$\tan\alpha = \frac{1}{2gx^2}\left\{2v_0^2 x \pm \sqrt{4v_0^4 x^2 - 4gx^2(2v_0^2 y + gx^2)}\right\}$$

We now impose the restriction

$$4v_0^4 x^2 - 4gx^2(2v_0^2 y + gx^2) = 0$$

since we want exactly one pair of x and y values to correspond to tan α and, since the tangent function is one-to-one on the domain of α, **if the discriminant = 0, there will be exactly one α corresponding to (x, y).** The condition

$$4v_0^4 x^2 - 4gx^2(2v_0^2 y + gx^2) = 0 \tag{2}$$

produces the needed relationship between x and y to produce the envelope curve. We write equation (2) in standard form to discover its shape and also to solve the volume phase of the problem.

$$v_0^4 = 2gv_0^2 y + g^2 x^2$$

$$y = \frac{v_0^2}{2g} - \frac{g}{2v_0^2} x^2 \tag{3}$$

Equation (3) is the equation for the envelope curve which is itself a parabola. This completes the first phase of the solution.

It is interesting to verify that for each α, equation (1) intersects the envelope equation, (3), in exactly one point and that (1) and (3) have a common tangent line at this point. Let y_T be a fountain trajectory and y_E be the envelope equation. Then $y_T = y_E$ implies that

$$\left(\frac{-gx^2}{2v_0^2}\right)(1 + \tan^2\alpha) + x\tan\alpha = \frac{v_0^2}{2g} - \frac{g}{2v_0^2} x^2$$

After an amount of algebraic prestidigitation this can be written as

$$(\tan^2\alpha)x^2 - (\tan\alpha)\left(\frac{2v_0^2}{g}\right)x + \frac{v_0^4}{g^2} = 0$$

Indeed, it is a perfect square trinomial!!

$$\left(x\tan\alpha - \frac{v_0^2}{g}\right)^2 = 0 \text{ implies } x = \frac{v_0^2}{g\tan\alpha}$$

This indicates exactly one x is common to both curves. To check that the general trajectory and the envelope have a common tangent line at their single common point we compute their derivatives and then let $x = \dfrac{v_0^2}{g \tan \alpha}$.

$$y_E' = -\dfrac{g}{v_0^2} x = \left(\dfrac{-g}{v_0^2}\right)\left(\dfrac{v_0^2}{g \tan \alpha}\right) = -\cot \alpha$$

$$y_T' = \left(\dfrac{-g x}{v_0^2}\right)(1+\tan^2 \alpha) + \tan \alpha = \dfrac{-g}{v_0^2}\left(\dfrac{v_0^2}{g \tan \alpha}\right)\sec^2 \alpha + \tan \alpha$$

$$y_T' = \dfrac{-\sec^2 \alpha + \tan^2 \alpha}{\tan \alpha} = \dfrac{-1}{\tan \alpha} = -\cot \alpha = y_E'$$

Now that there is absolutely no doubt that we have the envelope curve we proceed to phase (2). Rotating the area under the envelope about the y-axis we use the disk method to determine the volume.

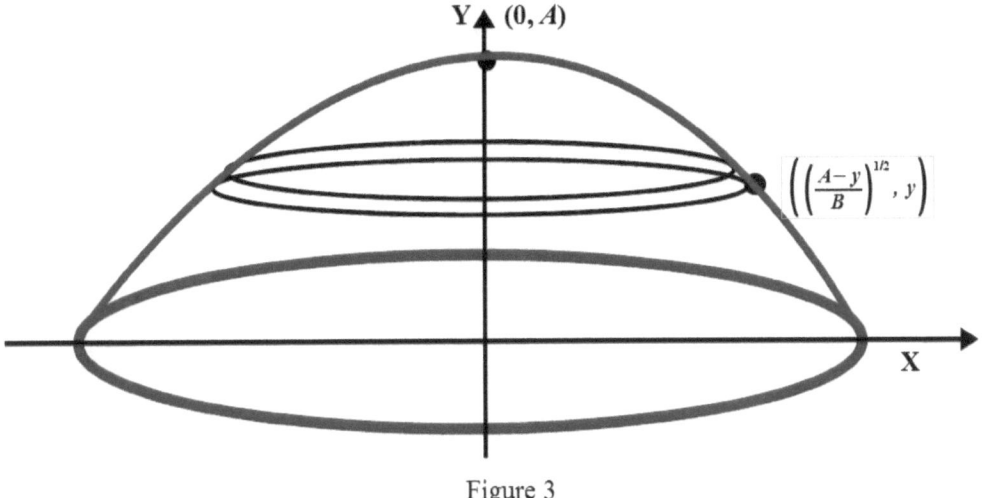

Figure 3

For convenience we write the envelope equation, (3) in the form

$$y = A - B x^2$$

where $A = \dfrac{v_0^2}{2g}$ and $B = \dfrac{g}{2 v_0^2}$. Since the thickness of the disk is dy, we solve for x to get the point on the right edge of the disk $\left(\left(\dfrac{A-y}{B}\right)^{1/2}, y\right)$ so that we see that the radius of the disk is $\left(\dfrac{A-y}{B}\right)^{1/2}$. We can set up the volume integral.

$$V = \pi \int_0^A \left(\frac{A-y}{B}\right) dy = \frac{\pi}{B}\left(Ay - \tfrac{1}{2}y^2\right)\Big|_0^A = \frac{\pi A^2}{2B}$$

Replacing A and B with the expressions in terms of v_0 and g we find

$$V = \frac{\pi v_0^6}{4g^3} = \left(\frac{\pi}{4g^3}\right) v_0^6$$

which means that the volume varies directly as the sixth power of the initial velocity of the nozzle jets!

The Volume of a Hypersphere

The hypersphere has the equation

$$x^2 + y^2 + x^2 + w^2 = R^2$$

if centered at the origin (0, 0, 0, 0) and has a radius of R in four dimensional space. We approach the project of determining its volume inductively by first considering volumes of "spheres" in lower dimensions. We start with the "sphere" of radius R in one dimension, centering it on the number line so that the center is at the origin, (0) as illustrated in Figure 1.

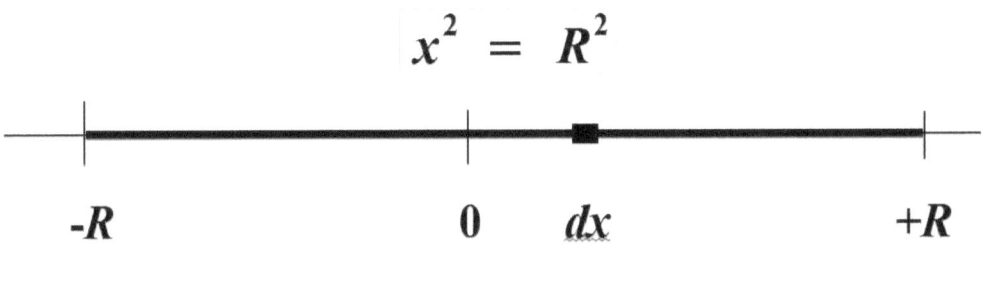

Figure 1

I. The Sphere in one dimension

We can find the "volume" (actually length in this case) of the "sphere", shown as the portion of the number line from $-R$ to R by integrating the *line segment* whose length is dx. Therefore

$$V_1 = \int_{-R}^{R} dx = 2R$$

II. The Sphere in Two Dimensions

The two dimensional "sphere" is really a circle. Proceeding inductively, we add one more integral to compute the "volume", which in this case is an area. Considering the equation

$$x^2 + y^2 = R^2,$$

we solve it for *y* to determine the extent of the sphere in the *y*-direction for a given value of *x* between –*R* and +*R*. Figure 2 shows the 2-D "sphere".

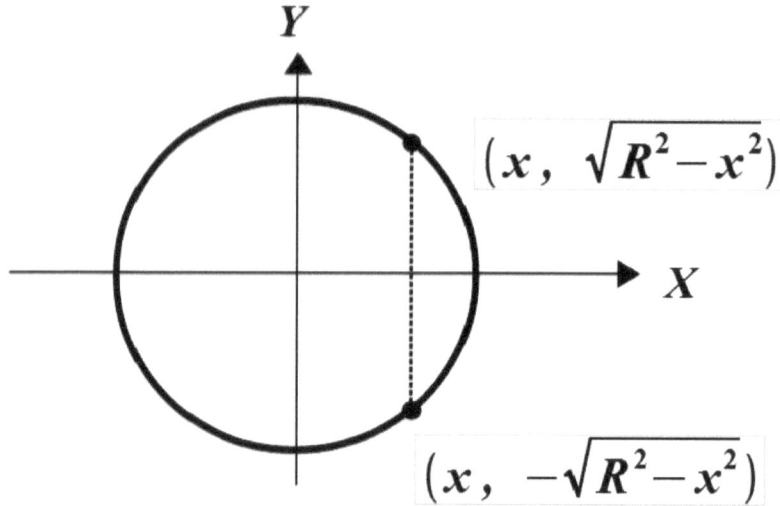

Figure 2

The only thing different here is the addition of the vertical axis. The "sphere" now extends from $-\sqrt{R^2-x^2}$ to $\sqrt{R^2-x^2}$ in the vertical direction as well as from –*R* to *R* in the horizontal direction as was the situation in case I. To obtain the "volume" we integrate a *square* whose dimensions are *dy* by *dx* as follows.

$$V_2 = \int_{-R}^{R} \int_{-\sqrt{R^2-x^2}}^{\sqrt{R^2-x^2}} dy\, dx = \int_{-R}^{R} 2\sqrt{R^2-x^2}\, dx = 2\left(\frac{1}{2}\pi R^2\right) = \pi R^2$$

We could have evaluated the double integral in polar coordinates more easily as shown below.

$$V_2 = \int_0^{2\pi} \int_0^{R} r\, dr\, d\theta = \int_0^{2\pi} \frac{1}{2} R^2\, d\theta = \pi R^2$$

III. The Three Dimensional Sphere

We give here the volume of the ordinary sphere computed with three integrals in order to show how to set up the 4-dimensional case with four integrals! The two integrals used in case II are retained. The third integral has limits specified by the extent of the sphere in the z-direction, determined by solving the sphere equation

$$x^2 + y^2 + z^2 = R^2$$

for z. To determine the extent of the sphere in the z-direction for a given pair of values (x,y), we solve for z to obtain $-\sqrt{R^2-(x^2+y^2)}$ and $\sqrt{R^2-(x^2+y^2)}$.

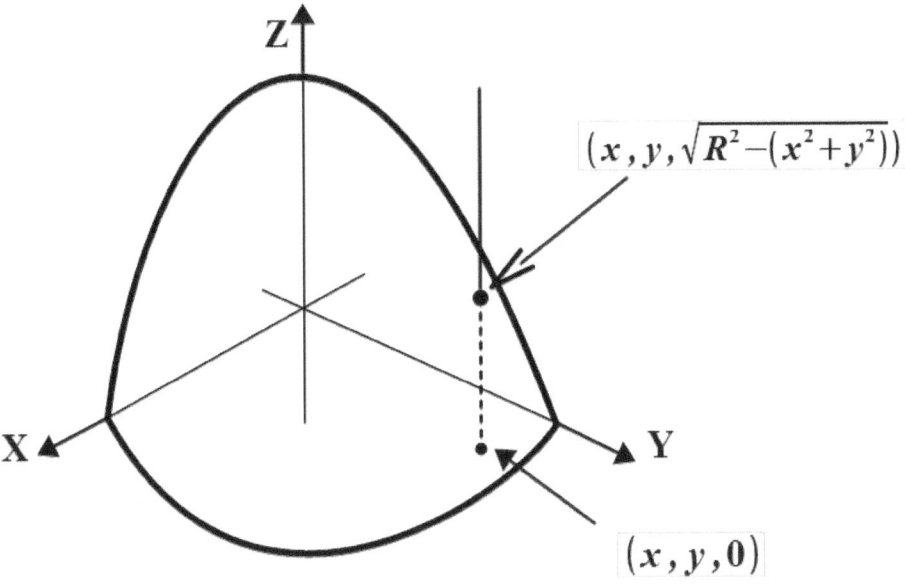

Figure 3

The Figure 3 shows the first octant of the sphere and the upper extent. Unseen is the lower extent at $(x, y, -\sqrt{R^2-(x^2+y^2)})$. These are the limits assigned to the third integral. We proceed to integrate the *cube* whose dimensions are *dx* by *dy* by *dz*. The triple integral set up is shown below.

$$V_3 = \int_{-R}^{R} \int_{-\sqrt{R^2-y^2}}^{\sqrt{R^2-y^2}} \int_{-\sqrt{R^2-(x^2+y^2)}}^{\sqrt{R^2-(x^2+y^2)}} dz\, dy\, dx$$

We begin to compute the volume by doing the *dz* integral and converting the *dx* and *dy* integrals to polar coordinates.

$$V_3 = 2\int_{-R}^{R} \int_{-\sqrt{R^2-x^2}}^{\sqrt{R^2-x^2}} \sqrt{R^2-(x^2+y^2)}\, dy\, dx$$

$$V_3 = 2\int_{0}^{2\pi} \int_{0}^{R} (\sqrt{R^2-r^2})r\, dr\, d\theta = \int_{0}^{2\pi} \left(\frac{2}{3}R^3\right)d\theta = \frac{4}{3}\pi R^3$$

IV. The Four Dimensional Hypersphere

At last we consider the volume V_4 for the hypersphere whose equation is

$$x^2+y^2+z^2+w^2 = R^2$$

This contains the w^2 term indicating that the hypersphere extends in another direction in space. This is a direction not specified by the *x*-direction, the *y*-direction, or the *z*-direction. To determine, analytically, the extent of the hypersphere in the *w*-direction for a given triple of values (*x,y,z*), we solve the hypersphere's equation for *w*. We get that *w* extends from $-\sqrt{R^2-(x^2+y^2+z^2)}$ to $\sqrt{R^2-(x^2+y^2+z^2)}$. These will be the limits of the fourth integral in the expression which will be used to calculate the volume of the hypersphere.

We integrate a ***hypercube*** of dimensions ***dx*** by ***dy*** by ***dz*** by ***dw*** by setting up the following four-integral expression.

$$V_4 = \int_{-R}^{R} \int_{-\sqrt{R^2-x^2}}^{\sqrt{R^2-x^2}} \int_{-\sqrt{R^2-(x^2+y^2)}}^{\sqrt{R^2-(x^2+y^2)}} \int_{-\sqrt{R^2-(x^2+y^2+z^2)}}^{\sqrt{R^2-(x^2+y^2+z^2)}} dw\, dz\, dy\, dx$$

The *dw* integral is easy to perform after which we utilize cylindrical coordinates, (r, θ, z), by converting to polar coordinates the region over which the first two integrals apply.

$$V_4 = 2 \int_{-R}^{R} \int_{-\sqrt{R^2-x^2}}^{\sqrt{R^2-x^2}} \int_{-\sqrt{R^2-(x^2+y^2)}}^{\sqrt{R^2-(x^2+y^2)}} \sqrt{R^2-(x^2+y^2+z^2)} \, dz \, dy \, dx$$

$$V_4 = 2 \int_{0}^{2\pi} \int_{0}^{R} \int_{-\sqrt{R^2-r^2}}^{\sqrt{R^2-r^2}} \sqrt{R^2-(r^2+z^2)} \, dz \, r \, dr \, d\theta$$

The inner integral, $\int_{-\sqrt{R^2-r^2}}^{\sqrt{R^2-r^2}} \sqrt{R^2-(r^2+z^2)} \, dz$, deserves special attention. With respect to z, r is a constant $\leq R$. Let $K^2 = (R^2 - r^2)$, so that K is not negative. Then the integral becomes

$$\int_{-\sqrt{R^2-r^2}}^{\sqrt{R^2-r^2}} \sqrt{R^2-(r^2+z^2)} \, dz = \int_{-K}^{K} \sqrt{K^2-z^2} \, dz = \frac{1}{2}\pi K^2 = \frac{1}{2}\pi(R^2-r^2)$$

as the second integral is the area under a semi-circle of radius K. Substituting this result, we continue computing the volume of the hypersphere.

$$V_4 = 2 \int_{0}^{2\pi} \int_{0}^{R} (\frac{1}{2}\pi(R^2-r^2)) r \, dr \, d\theta = \pi \int_{0}^{2\pi} (\int_{0}^{R} (R^2 r - r^3) \, dr) \, d\theta$$

$$V_4 = \pi \int_{0}^{2\pi} (\frac{1}{4} R^4) \, d\theta$$

$$V_4 = \frac{1}{2}\pi^2 R^4$$

Newton's Law of Heat Transfer and Measuring Very Hot Objects

Newton's Law of Heat Transfer is stated by

$$\frac{dR}{dt} = k(M-R) \qquad (1)$$

where R is the temperature of an object placed in a surrounding medium of constant temperature, M, and k is a constant. Equation (1) means that the rate of change in the temperature of the object, which is immersed in a medium, is proportional to the difference in the object's temperature and that of the surrounding medium such as air or a liquid.

It is desired to use a thermometer to measure very hot objects such as molten metals. Here a molten metal would itself be the surrounding medium of temperature M and $R(t)$ would be the temperature reading of the thermometer at time t elapsed from the time of immersion into the liquid metal. Since a typical thermometer would break due to a too sudden temperature change, the thermometer can be insulated so that the temperature changes would be sufficiently slow so that no damage to the instrument would occur. The trade off here would be a slow response instrument.

This article presents a method whereby an accurate reading can be determined *before* the insulated thermometer reaches the temperature of M! We can derive a formula based on three temperature readings of the instrument, the first being the "shelf" temperature of the thermometer before immersion. A challenging problem at this point could be "Plan a procedure and a formula which provides an accurate temperature of the medium being measured, with a well-insulated thermometer, in a short time."

Instead we provide the method here. Let R_0 be the shelf temperature at time $t = 0$, R_1 be the temperature at $t = t_1$, and R_2 be the temperature at $t = t_2$. We want these temperature observations to be equally spaced time-wise so that $t_1 - 0 = t_2 - t_1$. Thus the R_2 reading is made at $t = 2t_1$. We proceed to derive the formula.

Note that the difference in (1) is positive and that (1) is variable separable.

$$\int \frac{dR}{M-R} = k \int dt$$

$$-\ln(M-R) = kt + C \qquad (2)$$

To evaluate C we let $t = 0$ and $R = R_0$. Then we have

$$-\ln(M-R_0) = C.$$

Substituting into (2) and using the logarithm property $\ln(A) - \ln(B) = \ln(A/B)$ we obtain

$$\ln\left(\frac{M-R_0}{M-R}\right) = kt \qquad (3)$$

We evaluate k by substituting $R = R_1$ and $t = t_1$.

$$k = \frac{1}{t_1} \ln\left(\frac{M - R_0}{M - R_1}\right) \tag{4}$$

Substituting (4) into (3),

$$\ln\left(\frac{M - R_0}{M - R}\right) = \frac{t}{t_1} \ln\left(\frac{M - R_0}{M - R_1}\right) \tag{5}$$

In (5) we replace R with R_2 and t with $2t_1$. The result is

$$\ln\left(\frac{M - R_o}{M - R_2}\right) = \frac{2 t_1}{t_1} \ln\left(\frac{M - R_0}{M - R_1}\right)$$

Using the property that $P*\ln(x) = \ln(x^P)$ and then dropping the logarithms (by taking the e-based exponential of both sides) we have the result

$$\left(\frac{M - R_0}{M - R_2}\right) = \left(\frac{M - R_0}{M - R_1}\right)^2$$

which simplifies to

$$\frac{1}{M - R_2} = \frac{M - R_0}{(M - R_1)^2}.$$

Cross multiplying and simplifying yields the following.

$$(M - R_1)^2 = (M - R_0)(M - R_2)$$

$$M(2 R_1 - R_2 - R_0) = R_1^2 - R_2 R_0$$

$$M = \frac{R_1^2 - R_2 R_0}{2 R_1 - R_2 - R_0} \tag{6}$$

Equation (5) could be the final formula for the temperature of the medium, but we can do a further simplification with a little algebraic prestidigitation. We now express (6) in terms of temperature differences. Let $D_1 = R_1 - R_o$, $D_2 = R_2 - R_1$. Then the denominator of (6) has the representation

$$2 R_1 - R_2 - R_0 = -(R_2 - R_1) + (R_1 - R_0)$$

$$2 R_1 - R_2 - R_0 = D_1 - D_2.$$

The numerator requires a bit more work.

$$R_1^2 - R_2 R_0 = R_1^2 - R_0(D_2 + R_1)$$
$$= R_1^2 - R_0 R_1 - D_2 R_0$$
$$= R_1(R_1 - R_0) - D_2 R_0$$
$$= R_1 D_1 - D_2 R_0$$
$$= (D_1 + R_0) D_1 - D_2 R_0$$
$$= D_1^2 + R_0(D_1 - D_2)$$

We now have the final form of the formula

$$M = \frac{D_1^2 + R_0(D_1 - D_2)}{D_1 - D_2}$$

But wait! There is one last simplification.

$$M = R_0 + \frac{D_1^2}{D_1 - D_2}$$

This is yet another example in mathematics where persistence pays off.

A tiny microchip in an electronic version of the insulated thermometer could be programmed to measure the two differences and quite quickly produce the value of M way before the sensor actually reaches the temperature of the medium being measured.

Cooling Tea and Newton's Law of Heat Transfer

Introduction

The aim of this discussion is an analysis of a heat transfer problem. Such problems can often be seen in texts or in contests. Frequently these problems are incompletely stated or are open to so many interpretations that several different solutions can be developed based on one or more inferred assumptions. We will take as a case study the problem below which appeared on the Cambridge University Entrance Examination in November, 1975, with some editing.

> A cup of tea cools at a rate proportional to the difference between its temperature and that of its surroundings. In winter the room temperature is 15^0C and I must wait 10 minutes for my tea to cool from 90^0C to 50^0C. How long must I wait in summer when the room temperature is 25^0C?
> If I took milk in my tea, I could achieve part of the cooling by adding milk at 5^0C. So that I may drink the tea as soon as possible, should I add the milk just after pouring the tea, or just before drinking it?

Some Possible Questions and Assumptions

Clearly, if we add the milk initially, it will have a temperature of 5^0C. If we wait to add the milk until we are just about to consume the tea, was the milk out of the refrigerator during the time the tea was cooling, or do we take the milk from the refrigerator immediately before combining it with the tea? How much milk do we add anyway? This certainly would appear to matter! Does the amount added really matter, though? If we were to set both the 90^0C tea and 5^0C milk on a table and wait to add the milk, what are the heat flow assumptions? Do we assume they are in similar containers, or are there different heat flow characteristics for each container? Obviously, if the milk container were well insulated but the tea container were not, the milk would still be nearly at 5^0C when added later, and there would seem to be less of a difference between the add it now or add it later scenarios. Maybe not!

Analyzing and Solving the Problem

Let's analyze the problem sentence-by-sentence. The first sentence of the problem allows use of Newton's Law of Heat Transfer which in differential equation form is

$$\frac{dT}{dt} = k(T - M) \qquad (1)$$

where T is the temperature of the tea at time t, M is the constant temperature of the surrounding medium, and k is the constant of proportionality. The second sentence informs us that $M = 15^0C$ in the winter, and provides the necessary information to evaluate k and the integration constant, C. Next we are informed that 5^0C milk is to be added to the tea, either immediately after making the tea (when its temperature is 90^0C) or at the time it is sufficiently cool enough for drinking (we must assume that this is at 50^0C). The first question is straightforward, asking how long it takes for the tea to reach drinking temperature in the summer when M is changed to 25^0C. The second paragraph is where the problem grows more complex and subject to interpretation. (The original version of this problem has the milk at room temperature. The 5^0C temperature is more realistic.)

Answering the first question, we use the winter data and (1) to evaluate k which is affected by the insulating properties of the container for the tea.

$$\frac{dT}{dt} = k(T - 15)$$

$$\int \frac{dT}{T-15} = \int k\, dt$$

$$\ln(T - 15) = kt + C$$

We let $T = 90$ when $t = 0$, and $T = 50$ when $t = 10$ to evaluate both constants.

$$\ln(90 - 15) = C$$

$$\ln\left(\frac{T-15}{75}\right) = kt$$

$$k = \frac{1}{10}\ln\left(\frac{7}{15}\right)$$

Thus (1) can be written as

$$\frac{dT}{dt} = \frac{(T-15)}{10}\ln\left(\frac{7}{15}\right) \qquad (2)$$

Separating variables and integrating (2) yields

$$\ln(T - 15) = \frac{t}{10}\ln\left(\frac{7}{15}\right) + C \qquad (3)$$

which we now adapt to the summer season by replacing 15 with 25.

$$\ln(T - 25) = \frac{t}{10}\ln\left(\frac{7}{15}\right) + C \qquad (4)$$

Replacing T with 90 and t with 0 we find $C = \ln(65)$. Now (4) becomes

$$\ln\left(\frac{T-25}{65}\right) = \frac{t}{10}\ln\left(\frac{7}{15}\right)$$

Finally putting $T = 50$ and solving for t we find

$$\ln\left(\frac{25}{65}\right) = \frac{t}{10}\ln\left(\frac{7}{15}\right) \quad t = \frac{10\ln(5/13)}{\ln(7/15)} \approx 12.53722 \text{ minutes}$$

Let's now assume it is winter and we want to compare the cooling times for the tea to reach $50\,^0C$ when we either add the milk immediately, when the tea is at $90\,^0C$, or wait until later at which time we add the milk when the tea is just above $50\,^0C$ and then obtain a cup of tea at exactly $50\,^0C$. We assume that the amount of milk we are adding is 1/14 of the amount of tea since it does not require much milk to whiten and properly flavor a cup of tea. *For now we also assume that the milk is at $5\,^0C$ whenever it is added to the tea.*

If we add the milk immediately, we must accordingly adjust the value of T which corresponds to the time $t = 0$. This requires use of the First Law of Thermodynamics. For two liquids which are being combined this law states that the product of the mass, heat capacity, and change in temperature is the same for each liquid. Symbolically, if M_1 and M_2 are the two masses, H_1 and H_2 are the two heat capacities, T_1 and T_2 are the initial temperatures, and T_F is the final temperature of the combined liquids, we have

$$M_1 H_1 (T_1 - T_F) = M_2 H_2 (T_F - T_2)$$

where it is understood that $T_1 > T_F > T_2$. Since tea and milk are essentially water, we can safely assume $H_1 = H_2$ and the masses are proportional to the liquid volumes which we can call Q_1 and Q_2. Then the equation can be written as

$$Q_1(T_1 - T_F) = Q_2(T_F - T_2).$$

Solving for T_F we obtain

$$\frac{Q_1 T_1 + Q_2 T_2}{Q_1 + Q_2} = T_F \qquad (5)$$

which is equivalent to calculating a weighted average of the temperatures of the two liquids by volume! Since we have a tea to milk ratio of 14:1, we can obtain the starting temperature of the milk-flavored tea, T_0, at $t = 0$, as follows:

$$T_0 = \frac{(14)(90) + (1)(5)}{14+1} = 84.333^0$$

We can use the value of k obtained earlier for equation (2) as we assume the tea cools in the same container as before with the only difference being the cooler initial temperature. Now we proceed to employ equation (3) to calculate the time required for T to reach $50\,^0C$. We first obtain the integration constant C.

$$\ln(84.333-15) = \frac{0}{10}\ln\left(\frac{7}{15}\right) + C = C$$

$$C = \ln(69.3333)$$

Substituting for C we have

$$\ln\left(\frac{T-15}{69.3333}\right) = \frac{t}{10}\ln\left(\frac{7}{15}\right).$$

Letting $T = 50$ we solve for t.

$$t = \frac{10\ln\left(35/(69+\frac{1}{3})\right)}{\ln(7/15)} = \frac{10\ln(105/108)}{\ln(7/15)} \approx 8.969 \text{ minutes} \qquad (6)$$

Now we calculate the time needed to cool the tea without the milk to a temperature, just above 50^0C, at the precise time when the addition of the 5^0C milk reduces the temperature to 50^0C. We return to equation (3), this time evaluating C with the substitution of 90^0C for T when $t = 0$. Thus with $C = \ln(75)$ we have (3) leading to

$$\ln\left(\frac{T-15}{75}\right) = \frac{t}{10}\ln\left(\frac{7}{15}\right). \qquad (7)$$

We must solve (7) for T as a function of t and then work with the First Law of Thermodynamics to represent the temperature at time t when we add the milk at 5^0C to the tea at the temperature $T(t)^0C$. Solving (7) for T yields

$$T(t) = 15 + 75\left(\frac{7}{15}\right)^{t/10} \qquad (8)$$

Using the principle shown in equation (5) we have

$$\frac{14\left(15 + 75\left(\frac{7}{15}\right)^{t/10}\right) + (1)(5)}{14+1} = 50$$

$$\left(\frac{7}{15}\right)^{t/10} = \frac{107}{210}$$

$$t = \frac{10\ln(107/210)}{\ln(7/15)} \approx 8.847 \text{ minutes.}$$

We thus get slightly faster cooling by waiting to add the milk by about 7.33 *seconds*.

At this time let's assume that when we made the tea at 90^0C we took the milk out of the refrigerator at 5^0C and set it on the table beside the cup of tea in the room at 15^0C. Under this scenario, the milk is slowly warming while the tea is cooling. We make the further assumption that the tea and milk are in containers with similar insulation characteristics. This allows us to use the same constant in absolute value, $-k$, for the milk, with the minus giving a positive value for the growth constant since the temperature of the milk is an increasing function of time. Equation (8) already describes the temperature of the tea at time, t. We need a similar function describing the temperature of the milk as a function of t, which we call $T_M(t)$. We adapt equation (1) for the warming milk

$$\frac{dT_M}{dt} = -k(15 - T_M). \tag{9}$$

Note the reversal of the medium temperature and T_M since the derivative $\frac{dT_M}{dt}$ is positive. Of course, distributing the minus results in a differential equation identical to (1), but we wanted to stress the fact that the growth constant, $-k$, must be positive. Solving (9) subject to the condition that $T_M = 5$ when $t = 0$ we have

$$\int \frac{dT_M}{15 - T_M} = -\ln(15 - T_M) = -\frac{t}{10}\ln\left(\frac{7}{15}\right) + C.$$

Evaluating the boundary condition $T_M = 5$ when $t = 0$, we get $C = -\ln 10$ which produces the equation

$$\ln\left(\frac{10}{15 - T_M}\right) = \ln\left(\frac{7}{15}\right)$$

which we solve for $T_M(t)$.

$$\frac{15 - T_M}{10} = \left(\frac{7}{15}\right)^{t/10}$$

$$T_M(t) = 15 - 10\left(\frac{7}{15}\right)^{t/10} \tag{10}$$

We now apply formula (5) to the functions in equations (8) and (10) with $T_F = 50$. Thus we begin with

$$\frac{(14)(T(t)) + (1)(T_M(t))}{14 + 1} = 50$$

and then replace $T(t)$ and $T_M(t)$ with the expressions in equations (8) and (10).

$$\frac{(14)\left(15+75\left(\frac{7}{15}\right)^{t/10}\right)+(1)\left(15-10\left(\frac{7}{15}\right)^{t/10}\right)}{14+1}=50$$

$$1040\left(\frac{7}{15}\right)^{t/10}=525$$

$$\left(\frac{7}{15}\right)^{t/10}=\frac{105}{208}$$

$$t=\frac{10\ln(105/208)}{\ln(7/15)}\approx 8.9692\ minutes \tag{11}$$

Note the exact match between (6) and (11)! The conclusion is that it does not make any difference in the cooling time if the milk is added initially, or if the milk is left out to gradually warm while the tea cools and then is added just before the tea reaches 50°C.

Is This a Fluke?

The reader is entitled to think that perhaps the numbers have been rigged to cause the equal cooling times. Therefore we will prove that the equivalence must occur under the most general conditions for the following scenario. We are comparing the time for the tea to cool to a certain specified temperature for two cases. One case (Case I) involves adding the cold milk to the tea initially, and then letting it cool to the specified temperature. The other case (Case II) allows the temperatures of both the cold milk and the hot tea to move toward the temperature of the surrounding medium separately, after which they are combined at the instant the tea-milk mix would have the specified temperature.

We define certain constants before analyzing Case I and Case II.
Let G = the initial tea temperature before any milk is added at time $t = 0$
M = the temperature of the surrounding medium (the air temperature)
t_D = the time for the tea to reach the temperature of D degrees, the drinkable temperature
A = the temperature of the milk when it is taken from the refrigerator ($t = 0$)
Q_1 = the quantity of tea before the milk is added
Q_2 = the quantity of milk
k = the proportionality constant in Newton's Law of Heat Transfer

For the temperatures we have $A < M < D < G$. First we determine the value of k in terms of the other constant parameters. We begin with

$$\frac{dT_{TEA}}{dt}=k(T_{TEA}-M)$$

using the data points $T_{TEA} = G$ when $t = 0$, and $T_{TEA} = D$ when $t = t_D$. Separating the variables T_{TEA} and t, and integrating,

$$\int \frac{dT_{TEA}}{T_{TEA}-M} = \int k\,dt$$

$$\ln(T_{TEA}-M) = kt + C$$

Substituting the first data point yields $C = \ln(G - M)$ so that

$$\ln\left(\frac{T_{TEA}-M}{G-M}\right) = kt.$$

Substituting the second data point and isolating k we find

$$\frac{1}{t_D}\ln\left(\frac{D-M}{G-M}\right) = k \tag{12}$$

Since we are assuming that the containers of tea and milk have the same heat-transfer characteristics, this value of k will be used in both Case I and Case II.

CASE I: Combine the Tea and Milk Initially

Using the principle shown in equation (5) we compute $T_{TM,0}$, the temperature of the tea-milk mix at time $t = 0$.

$$T_{TM,0} = \frac{Q_1 G + Q_2 A}{Q_1 + Q_2} \tag{13}$$

We now have the equation

$$\ln(T_{TM}-M) = kt + C$$

which relates the tea-milk temperature to the time, into which we insert the k-value computed above to obtain

$$\ln(T_{TM}-M) = \frac{t}{t_D}\ln\left(\frac{D-M}{G-M}\right) + C \tag{14}$$

Since $t = 0$ results in $T_{TM} = T_{TM,0}$ as computed in (13), we get

$$C = \ln\left(\frac{Q_1 G + Q_2 A}{Q_1 + Q_2} - M\right)$$

so that the relationship between T_{TM} and t is given by

$$\ln\left(\frac{T_{TM} - M}{\frac{Q_1 G + Q_2 A}{Q_1 + Q_2} - M}\right) = \frac{t}{t_D} \ln\left(\frac{D - M}{G - M}\right).$$

Now we substitute D for T_{TM}, and solve for the time, t, for the tea-milk mixture to arrive at the desired temperature.

$$t = (t_D)\left(\frac{\ln\left(\frac{D-M}{\frac{Q_1 G + Q_2 A}{Q_1 + Q_2} - M}\right)}{\ln\left(\frac{D-M}{G-M}\right)}\right) \tag{15}$$

CASE II: Let the Tea and Milk Cool Separately, Then Combine

We refer back to equation (14) in the form

$$\ln(T_{TEA} - M) = \frac{t}{t_D} \ln\left(\frac{D-M}{G-M}\right) + C$$

and use the data point $T_{TEA} = G$ when $t = 0$ to obtain the value of $C = \ln(G - M)$. Then after substituting and solving for T_{TEA} we have

$$T_{TEA}(t) = M + (G - M)\left(\frac{D-M}{G-M}\right)^{t/t_D} \tag{16}$$

which we temporarily put aside until we get an expression for the temperature of the milk, T_{MILK}, as a function of t. We begin with equation (1) and we take the *opposite* of the expression in (12) for the value of k since the temperature of the milk rises.

$$\frac{dT_{MILK}}{dt} = \frac{-1}{t_D} \ln\left(\frac{D-M}{G-M}\right)(M - T_{MILK})$$

Note that the factor $(M - T_{MILK}) > 0$ ensuring that $\frac{dT_{MILK}}{dt} > 0$. Separating the variables and integrating we obtain

$$-\ln(M - T_{MILK}) = \frac{-t}{t_D} \ln\left(\frac{D-M}{G-M}\right)(M - T_{MILK}) + C.$$

For the milk we have the data point $T_{MILK} = A$ when $t = 0$, leading to $C = -\ln(M - A)$. Substituting, deleting minus signs, and compacting the logarithms results in

$$\ln\left(\frac{M-T_{MILK}}{M-A}\right)=\ln\left(\frac{D-M}{G-M}\right)^{t/t_D}.$$

Dropping logarithms and solving for the temperature of the milk as a function of t we have

$$T_{MILK}(t)=M-(M-A)\left(\frac{D-M}{G-M}\right)^{t/t_D} \tag{17}$$

We use formula (5) to determine the temperature of the mixture of tea and milk as a function of t by substituting expressions (16) and (17) and set the result equal to the desired drinking temperature, D. Solving for t will give the required time for cooling.

$$\frac{(Q_1)(T_{TEA})+(Q_2)(T_{MILK})}{Q_1+Q_2}=D$$

The next few moves involve a considerable amount of algebraic manipulation.

$$Q_1\left(M+(G-M)\left(\frac{D-M}{G-M}\right)^{t/t_D}\right)+Q_2\left(M-(M-A)\left(\frac{D-M}{G-M}\right)^{t/t_D}\right)=D(Q_1+Q_2)$$

$$\{Q_1G+Q_2A-M(Q_1+Q_2)\}*\left(\frac{D-M}{G-M}\right)^{t/t_D}=(Q_1+Q_2)(D-M)$$

$$\left(\frac{D-M}{G-M}\right)^{t/t_D}=\frac{(Q_1+Q_2)(D-M)}{Q_1G+Q_2A-M(Q_1+Q_2)}=\frac{D-M}{\frac{Q_{1G}+Q_{2A}}{Q_1+Q_2}-M}$$

$$\left(\frac{t}{t_D}\right)\ln\left(\frac{D-M}{G-M}\right)=\ln\left(\frac{D-M}{\frac{Q_1G+Q_2A}{Q_1+Q_2}-M}\right)$$

$$t=(t_D)\frac{\ln\left(\frac{D-M}{\frac{Q_1G+Q_2A}{Q_1+Q_2}-M}\right)}{\ln\left(\frac{D-M}{G-M}\right)}$$

Since this is the same expression obtained in equation (15), Case I, we have proved that the temperature of D can be reached in the same time whether or not we mix the milk with the tea. This requires, however, that in the event we do not add the milk immediately, we leave the milk out of the refrigerator in the same environment shared by the container of tea. If the milk were at room temperature at all times as was the case in the original version of the Cambridge University Entrance Examination, we only need to replace A with M. The two times in Case I and Case II would still be equal.

Conclusion

What made the cooling tea problem not straightforward was the fact that the different scenarios were possible within the context of the problem statement. There was a genuine difference between the two situations: (1) the milk was maintained at a constant temperature, and (2) the milk was allowed to warm toward the room temperature. With the particular data provided in the problem statement, the second situation resulted in equal cooling times regardless of whether the milk was immediately added or added at the end of the cooling period for the tea. It was interesting to check that this result was not merely an artifact of the numbers, but was ***generally*** true for arbitrary initial conditions and environment temperature as long as the two containers had identical heat-transfer characteristics.

Kepler's Laws: The Radius Vector Sweeps Out Equal Areas in Equal Time Increments, and All Orbits are Conic Sections

We begin by setting up a reference frame for the planet moving along the polar coordinate curve $r = f(\theta)$. We consider two perpendicular unit vectors u_θ and u_r as shown below in Figure 1. (Bold letters usually signify vectors.)

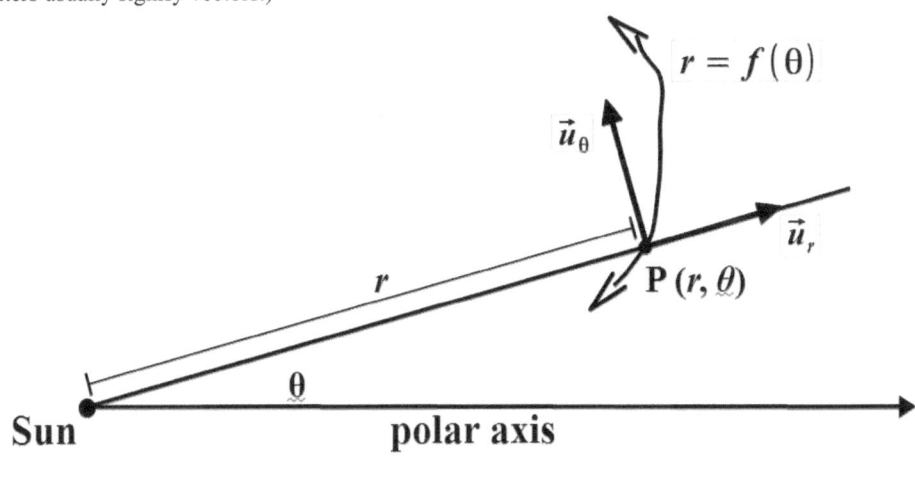

Figure 1

The planet at the point $P(r, \theta)$ moves along the curve $r = f(\theta)$ under the force of gravity with the sun at the origin. The unit vectors are given by

$$u_r = i\cos(\theta) + j\sin(\theta)$$
$$u_\theta = -i\sin(\theta) + j\cos(\theta).$$

We seek the velocity and acceleration vectors for the moving point P relative to the sun fixed at the origin. We begin with the position vector (the left side r is a vector function and the right side r is the $r = f(\theta)$ curve seen in Figure 1),

$$r(t) = r(t)u_r,$$

and proceed to obtain vectors v and a from $r(t)$. Since this will involve the derivatives of the unit vectors u_θ and u_r we note the following:

$$\frac{du_r}{d\theta} = -i\sin(\theta) + j\cos(\theta) = u_\theta \tag{1a}$$

$$\frac{du_\theta}{d\theta} = -i\cos(\theta) - j\sin(\theta) = -u_r \qquad (1b)$$

We now consider the curve $r = f(\theta)$ to be parameterized as follows:

$$r = r(t), \quad \theta = \theta(t), \quad t = \text{time, the independent variable.}$$

Since the derivative of the position vector with respect to time produces the velocity vector $v(t)$ we have

$$v(t) = \frac{dr(t)}{dt} = D_t\left(r(t) * u_r\right)$$

$$= \frac{dr}{dt} u_r + r \frac{du_r}{dt}$$

Using the chain rule on the second term we have

$$v(t) = \frac{dr}{dt} u_r + r * \frac{du_r}{d\theta} * \frac{d\theta}{dt}$$

Using (1 a and b) we get

$$v(t) = \frac{dr}{dt} u_r + r * \frac{d\theta}{dt} * u_\theta \qquad (2)$$

This states $v(t)$ in terms of its components in the u_r and u_θ directions. Since $a(t) = \frac{dv}{dt}$, we proceed to differentiate (2) with respect to t noting that the triple product rule must be used on the second term.

$$a(t) = D_t\left(\frac{dr}{dt} u_r + r * \frac{d\theta}{dt} * u_\theta\right)$$

$$a(t) = \left(\frac{d^2 r}{dt^2} * u_r + \frac{dr}{dt} * \frac{du_r}{dt}\right) +$$

$$[\frac{dr}{dt} * \frac{d\theta}{dt} * u_\theta + r * \frac{d^2\theta}{dt^2} * u_\theta + r * \frac{d\theta}{dt} * \frac{du_\theta}{dt}]$$

There are two substitutions to make, both using the chain rule.

$$\frac{du_r}{dt} = \frac{du_r}{d\theta} * \frac{d\theta}{dt} = \frac{d\theta}{dt} * u_\theta$$

$$\frac{du_\theta}{dt} = \frac{du_\theta}{d\theta} * \frac{d\theta}{dt} = -\frac{d\theta}{dt} * u_r$$

Placing these representations into the expression for the acceleration vector yields

$$a = \frac{d^2 r}{dt^2} * u_r + \frac{dr}{dt} * \frac{d\theta}{dt} u_\theta + \frac{dr}{dt} * \frac{d\theta}{dt} u_\theta$$
$$+ r * \frac{d^2\theta}{dt^2} * u_\theta - r\left(\frac{d\theta}{dt}\right)^2 u_r$$

Rearranging terms we have the final version.

$$a = \left[\frac{d^2 r}{dt^2} - r\left(\frac{d\theta}{dt}\right)^2\right] u_r + \left[r * \frac{d^2\theta}{dt^2} + 2 * \frac{dr}{dt} * \frac{d\theta}{dt}\right] u_\theta$$

Thus we have the acceleration in terms of the u_θ and u_r components which we list below.

$$a_r = \frac{d^2 r}{dt^2} - r\left(\frac{d\theta}{dt}\right)^2 \qquad (3)$$

$$a_\theta = r * \frac{d^2\theta}{dt^2} + 2 * \frac{dr}{dt} * \frac{d\theta}{dt} \qquad (4)$$

There is a much better way to express (4). We multiply and divide by r and think of the product rule for the derivative.

$$a_\theta = \left(\frac{1}{r}\right)\left(r^2 * \frac{d^2\theta}{dt^2} + 2r * \frac{dr}{dt} * \frac{d\theta}{dt}\right) \qquad (5a)$$

$$a_\theta = \frac{1}{r} * \frac{d}{dt}\left(r^2 d\theta\right) \qquad (5b)$$

We can derive two of Kepler's Laws from equations (3) and (5 a & b).

The Radius Vector Sweeps Out Equal Areas in Equal Time Increments

Since gravity produces a central force field, we can think of all the force of gravity as being directed toward the sun which is in the r-direction. There is zero force in the θ-direction. Looking at equations (5 a and b), since $r \neq 0$, we see that

$$\frac{1}{r} * \frac{d}{dt}\left(r^2 \frac{d\theta}{dt}\right) = 0$$

$$\text{implies } \frac{d}{dt}\left(r^2 \frac{d\theta}{dt}\right) = 0$$

If the derivative of a function is zero, that function is a constant function. Therefore

$$r^2\frac{d\theta}{dt} = h = \text{constant} \tag{6}$$

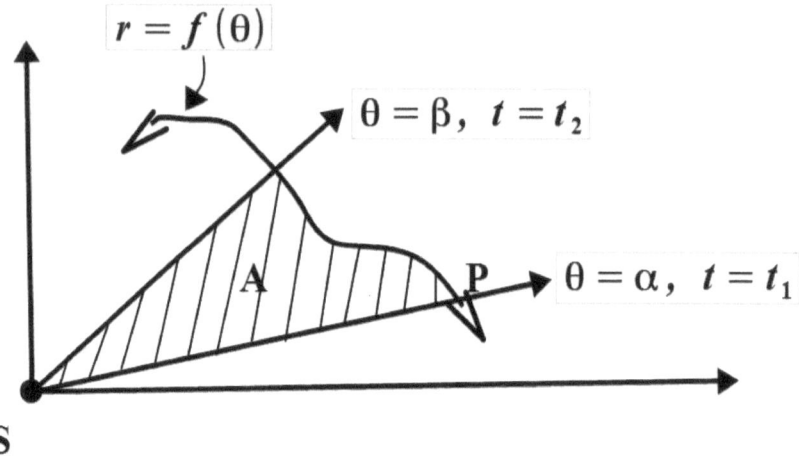

Figure 2 Polar Coordinate Area

The process of determining the area bounded as shown in Figure 2 with the outer boundary $r = f(\theta)$ is with the formula

$$A = \int_\alpha^\beta \frac{1}{2}r^2 d\theta$$

In view of equation (6) we get

$$A = \int_\alpha^\beta \frac{1}{2}h\, d\theta = \int_{t_1}^{t_2} h\, dt$$

$$= \frac{1}{2}h(t_2 - t_1) = \frac{1}{2}h\Delta t$$

Consider two time intervals $I_1 = [t_1,\ t_2]$ and $I_2 = [t_3,\ t_4]$ where $\Delta t = t_2 - t_1 = t_4 - t_3$. The radius vector sweeps out the areas A_1 and A_2, as follows:

$$A_{I_1} = \frac{1}{2}h(t_2-t_1) = \frac{1}{2}h(t_4-t_3) = A_{I_2}$$

That is, equal areas are swept out by the radius vector in equal time increments.

All Orbits are Conic Sections

From Newton's Law, $F = ma$ and the Universal Law of Gravity, $F = \dfrac{GMm}{r^2}$ we can determine the equation of the path of a planet of mass m orbiting a sun of mass M. Since gravity is a central force field, all the force is in the u_r direction so that there is zero force in the u_θ direction. Equating the two expressions for F and canceling m we have

$$a_r = -\frac{GM}{r^2}.$$

From equation (3) we have

$$\frac{d^2 r}{dt^2} - r\left(\frac{d\theta}{dt}\right)^2 = -\frac{GM}{r^2}$$

and from equation (6) we can replace $\dfrac{d\theta}{dt}$ with $\dfrac{h}{r^2}$. The result is the differential equation

$$\frac{d^2 r}{dt^2} - \frac{h^2}{r^3} = -\frac{GM}{r^2}. \tag{7}$$

The trick is to solve this equation in such a way that only r and θ appear. The key is the substitution

$$r = \frac{1}{z}. \tag{8}$$

This requires the replacement of $\dfrac{d^2 r}{dt^2}$ with an expression lacking the t. This needs careful use of the chain rule.

$$\frac{dr}{dt} = D_t\left(\frac{1}{z}\right) = \frac{-1}{z^2} * \frac{dz}{dt} = -\frac{1}{z^2} * \frac{dz}{d\theta} * \frac{d\theta}{dt}$$

$$= \frac{-1}{z^2} * \frac{dz}{d\theta} * \frac{h}{r^2} = -h\frac{dz}{d\theta}$$

$$\frac{d^2 r}{dt^2} = D_t\left(-h\frac{dz}{d\theta}\right) = \frac{d}{d\theta}\left(-h\frac{dz}{d\theta}\right)\frac{d\theta}{dt}$$

$$= -h * \frac{d^2 z}{d\theta^2} * \frac{h}{r^2} = -h^2 z^2 \frac{d^2 z}{d\theta^2}$$

The last expression for the second derivative along with (8) is substituted into equation (7) to obtain

or
$$-h^2 z^2 \frac{d^2 z}{d\theta^2} - h^2 z^3 = -GM z^2$$

$$\frac{d^2 z}{d\theta^2} + z = \frac{GM}{h^2}.$$

For convenience let
$$w = z - \frac{GM}{h^2}$$

so that $\frac{d^2 z}{d\theta^2} = \frac{d^2 w}{d\theta^2}$ and the above differential equation is replaced with the more streamlined equation

$$\frac{d^2 w}{d\theta^2} + w = 0. \tag{9}$$

This is a second order differential equation with the independent variable, θ, missing. The standard approach is the following. Let $u = dw/d\theta$ so that

$$\frac{du}{d\theta} = \frac{du}{dw} * \frac{dw}{d\theta} = u\frac{du}{dw} = \frac{d^2 w}{d\theta^2}$$

Then (9) becomes
$$u\frac{du}{dw} + w = 0.$$

This is easily solved. We begin by multiplying to obtain $u\,du + w\,dw$ which after integrating becomes

$$\frac{1}{2}u^2 + \frac{1}{2}w^2 = \frac{1}{2}C^2$$

or
$$u = \pm\sqrt{C^2 - w^2} = \frac{dw}{d\theta}$$

Integrating again yields

$$\int \frac{dw}{\sqrt{C^2 - w^2}} = \int \frac{\frac{1}{C}dw}{\sqrt{1 - \left(\frac{w}{C}\right)^2}} = \pm \int d\theta$$

$$\sin^{-1}\left(\frac{w}{C}\right) = C_1 \pm \theta = \pm(\theta + C_1)$$

$$w = C\sin(\pm(\theta + C_1))$$
$$w = -C\sin(\theta + C_1).$$

Note how we let the constants "swallow" the plus or minus sign. Looking back to $w = z - \frac{GM}{h^2}$ we have

$$z = \frac{GM}{h^2} - C\sin(\theta + C_1)$$

Factoring out GM/h^2 we have

$$z = \frac{GM}{h^2}\left(1 - \frac{h^2 C}{GM} * \sin(\theta + C_1)\right)$$

By equation (8) we obtain

$$r = \frac{h^2/GM}{1 - (h^2 C/GM)\sin(\theta + C)}$$

Let $e = \frac{h^2 C}{GM}$ so our equation becomes

$$r = \frac{e/C}{1 - e * \sin(\theta + C_1)} \qquad (10)$$

This is a polar form for a conic section, the nature of which depends on the range of values of e. Consider the notation in Figure 3 to see how this describes a conic section.

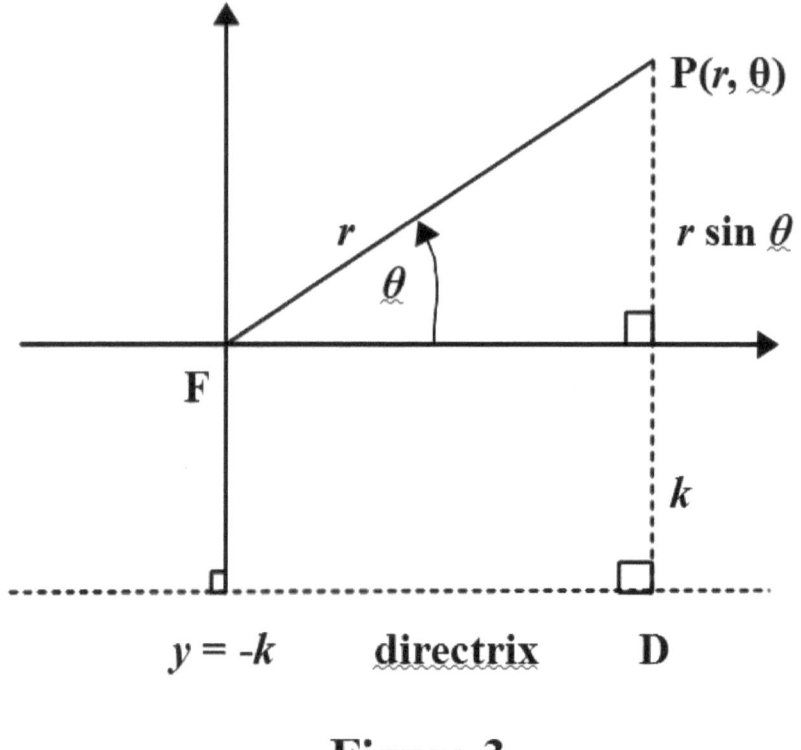

Figure 3

The definition of the eccentricity, e, of a conic section is the ratio $e = \dfrac{PF}{PD}$ where PD is the distance from point P to the directrix and PF is the distance from point P to the focus at F. This leads to

$$e = \frac{r}{1 - r\sin\theta}$$

which when solved for r becomes

$$r = \frac{ke}{1 - e\sin\theta}$$

which is an ellipse if $0 < e < 1$, a parabola if $e = 1$, and a hyperbola if $e > 1$. Comparing this with equation (10) we have shown that a planetary orbit is a conic section with the sun at a focus. Equation (10) is a version of this equation with the axes rotated if C_1 is nonzero.

The Derivation of a Pursuit Curve

Introduction

A heat-seeking missile always moves directly toward its target. It focuses only on the heat emitted by the target, never aiming for a point ahead of the target. The path generated by this movement is called a pursuit curve. The idea seems simple enough, but the actual mathematical analysis of such a curve requires a bit of fancy footwork. This is why you do not see such curves discussed in ordinary textbooks.

In this article we consider a pursuit curve in the context of a fox chasing a rabbit. The fox acts exactly like a heat-seeking missile. Its fixed gaze on the rabbit takes the place of the infrared sensors, causing the fox to maintain a constant heading directly toward the unfortunate rabbit.

Let's Cut to the Chase

Suppose a fox is initially located (at time $t = 0$) b units north of a rabbit which is initially at the origin. The rabbit moves east at a constant rate of r while at all times the fox moves directly toward the rabbit at the constant rate of f. Given that $f > r$, where will the fox intercept the rabbit?

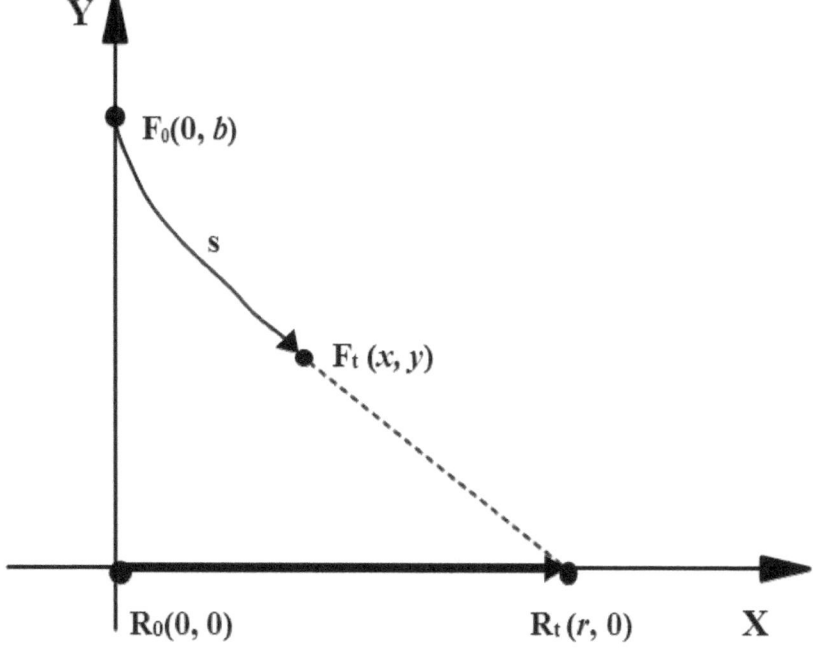

Figure 1

Figure 1 shows the configuration of the situation with a coordinate system imposed with the rabbit initially at the origin with the fox on the positive y-axis. Let s be the distance traveled by the fox along its path from F_0 to F_t. Then $\dfrac{ds}{dt} = f$. Also since the fox continually aims directly toward the rabbit, $\dfrac{dy}{dx} = \dfrac{y}{x-rt}$. We also know that when $t = 0$, $x = 0$, $y = b$, and $\dfrac{dx}{dy} = 0$ since the fox's path is initially tangent to the y axis. $ds = \sqrt{dx^2 + dy^2}$ so that $\dfrac{ds}{dy} = -\sqrt{1 + \left(\dfrac{dx}{dy}\right)^2}$ since s decreases as y increases.

Using the chain rule, we have $\dfrac{ds}{dt} = f = \dfrac{ds}{dy} * \dfrac{dy}{dt}$ which implies that

$$f = \dfrac{-dy}{dt}\sqrt{1 + \left(\dfrac{dx}{dy}\right)^2} \qquad (1)$$

To accommodate expression (1) for f we employ the inverted slope expression $\dfrac{dx}{dy} = \dfrac{x-rt}{y}$. If we were to substitute this into (1) now we would create a monster! Watch this cute maneuver! Let $p = \dfrac{dx}{dy}$. Then we have from $p = \dfrac{x-rt}{y}$ two items:

$$f = -\dfrac{dy}{dt}\sqrt{1 + p^2} \qquad (2)$$

$$p\,y = x - rt \qquad (3)$$

Differentiating with respect to y we see that

$$p + y\dfrac{dp}{dy} = \dfrac{dx}{dy} - r\dfrac{dt}{dy}$$

$$p + y\dfrac{dp}{dy} = p - r\dfrac{dt}{dy}$$

$$\dfrac{-y}{r}\dfrac{dp}{dy} = \dfrac{dt}{dy}$$

$$\dfrac{dy}{dt} = -\dfrac{r}{y}\dfrac{dy}{dp}$$

Substituting this last expression into (2) yields

$$f = -\left(\dfrac{-r}{y}\dfrac{dy}{dp}\right)\sqrt{1 + p^2}$$

$$\frac{dp}{dy} = \frac{1}{y} * \frac{r}{f} * \sqrt{1+p^2}$$

The last expression is a variable separable differential equation which we eagerly grasp for further processing. Getting the p's and y's separated we have

$$\int \frac{dp}{1+p^2} = \frac{r}{f} \int \frac{dy}{y}.$$

For convenience we lump $\frac{r}{f}$ by letting $k = \frac{r}{f}$. Clearly since $0 < r < f$, dividing through by f we see that $0 < k < 1$. Now we are solving

$$\int \frac{dp}{1+p^2} = k \int \frac{dy}{y}. \tag{4}$$

Let $\tan(\Phi) = p$. Then we draw the reference triangle in Figure 2 to be able to obtain any other needed trig function of Φ.

$$dp = \sec^2 \Phi \, d\Phi \text{ and } v^2 + 1 = \sec^2 \Phi$$

We substitute into the integral on the left of (4) and proceed with the solution.

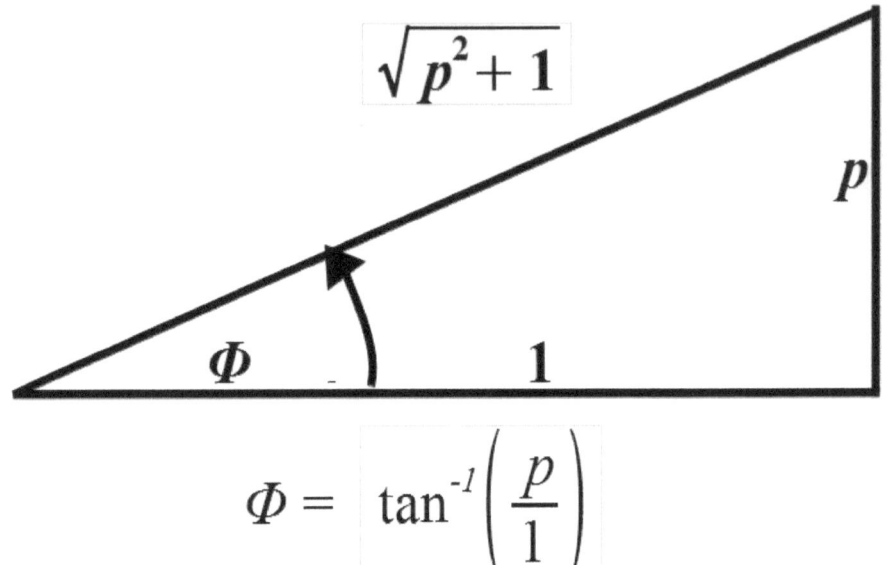

Figure 2

$$\int \frac{\sec^2 \Phi \, d\Phi}{\sec \Phi} = k \int \frac{dy}{y}$$

$$\int \sec \Phi \, d\Phi = k \int \frac{dy}{y}$$

$$\ln|\sec \Phi + \tan \Phi| = k(\ln|y| + \ln C) = k \ln|Cy| = \ln|(Cy)^k|$$

Substituting from the reference triangle and taking the exponential of both sides we find

$$\sqrt{1+p^2} + p = (Cy)^k$$
$$\left(\sqrt{1+p^2}\right)^2 = \left((Cy)^k - p\right)^2$$

We evaluate the constant of integration, C, with the condition that if $y = b$, then $p = \frac{dx}{dy} = 0$ so $1 = (Cb)^k$ implies that $C = 1/b.$ Substituting for C,

$$\left(\sqrt{1+p^2}\right)^2 = \left(\left(\frac{y}{b}\right)^k - p\right)^2$$

Squaring and simplifying we have

$$1 = \left(\frac{y}{b}\right)^{2k} - 2p\left(\frac{y}{b}\right)^k$$

which is readily solved for p.

$$2p\left(\frac{y}{b}\right)^k = \left(\frac{y}{b}\right)^{2k} - 1$$

$$p = \frac{1}{2}\left(\left(\frac{y}{b}\right)^k - \left(\frac{y}{b}\right)^{-k}\right) = \frac{dx}{dy}$$

We can isolate dx and then obtain x as a function of y for the pursuit curve.

$$dx = \frac{1}{2}\left(\left(\frac{y}{b}\right)^k - \left(\frac{y}{b}\right)^{-k}\right)dy = \frac{1}{2}\left(\frac{1}{b^k}y^k - b^k y^{-k}\right)dy$$

$$x = \frac{1}{2}\int\left(\frac{1}{b^k}y^k - b^k y^{-k}\right)dy = \frac{1}{2}\left\{\frac{1}{b^k}*\frac{1}{k+1}y^{k+1} - \frac{b^k}{1-k}y^{1-k}\right\} + C$$

Since the fox catches the rabbit on the x-axis, the x-intercept of this pursuit curve solves the problem. To get the x-intercept we let $y = 0$ obtaining $x = C$. Therefore the value of C tells us how far from the origin the rabbit meets its fate. The data point to use in evaluating C is the fox's initial position $(x, y) = (0, b)$.

$$0 = \frac{1}{2}\left\{\frac{1}{b^k}*\frac{1}{k+1}*b^{k+1} + b^k*\frac{1}{k-1}*b^{1-k}\right\} + C$$

$$0 = \frac{1}{2}\left\{\frac{b}{k+1} + \frac{b}{k-1}\right\} + C = \frac{1}{2}\left\{\frac{bk - b + bk + b}{k^2 - 1}\right\} + C$$

$$C = \frac{bk}{1-k^2}$$

Since earlier we let

$$k = \frac{r}{f} = \frac{rabbit's\ speed}{fox's\ speed}$$

we can write

$$C = \frac{b\left(\frac{r}{f}\right)}{1 - \frac{r^2}{f^2}} * \left(\frac{f^2}{f^2}\right)$$

or

$$C = \frac{brf}{f^2 - r^2}$$

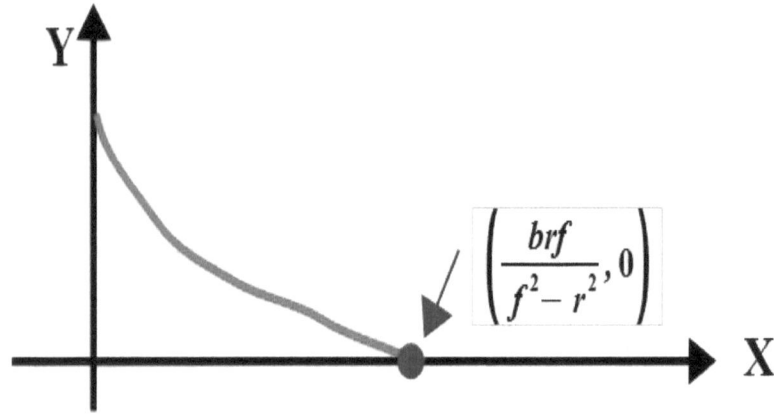

Figure 3

Notice that if the value of f is very close to the value of r, C becomes very large. Figure 3 shows the meeting of the fox with the rabbit.

Author's Note: While presenting the pursuit curve to a class he wondered how fast the fox must travel in order to intercept the rabbit at $(b, 0)$. This would be at such a speed that the rabbit would travel the same distance to the interception point that it was from the fox initially. Let $f = \Phi r$. Then

$$\frac{b\Phi r^2}{\Phi^2 r^2 - r^2} = b \quad \Rightarrow \quad \Phi^2 - \Phi - 1 = 0 \quad \Rightarrow \quad \Phi = \frac{1 + \sqrt{5}}{2} \approx 1.618.$$

This is the Golden Ratio! A survey of the literature revealed no such observation, so this result may not be known, or at least not widely known.

Optimizing a Storage Area

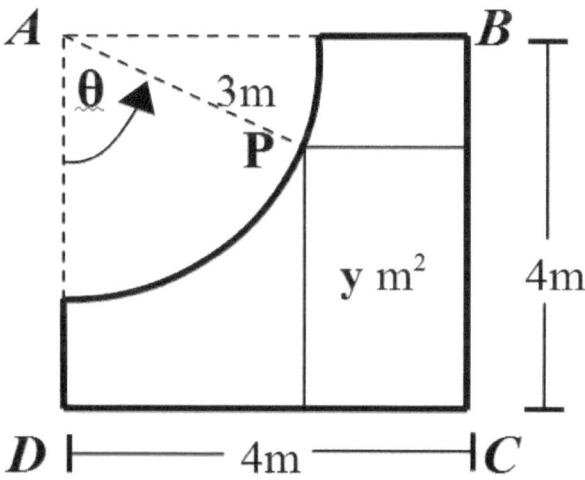

Figure 1

The diagram shows the plan of a store room. The plan consists of a square *ABCD* of side 4*m* from which a quadrant of a circle with center *A* and radius 3*m* has been removed. The owner intends to store a rectangular crate with one corner of its base at point *C*, and the opposite corner *P* of the base against the curved wall, where angle $DAB = \theta$ and $0 \leq \theta \leq \pi/2$. The base of the crate has area y m².

(i) Show that

$$\frac{dy}{d\theta} = 3(\sin\theta - \cos\theta)(4 - 3\sin\theta - 3\cos\theta),$$

and find the values of θ for which $dy/d\theta = 0$.

(ii) Determine the critical values of θ which produce a minimum value of *y*, and which produce a maximum value of *y*. Determine the absolute maximum and minimum values of the area, *y*.

Solution:
(I) Study Figure 2. From right triangle AGP we know that GP = DF = 3 sin θ and AG = BE = 3 cos θ. Thus FC = 4 – 3 sin θ and EC = 4 – 3 cos θ. Clearly the area is given by $y = (4 - 3 \sin\theta)(4 - 3 \cos\theta)$ with θ in the interval [0, π/2]. Thus 0 and π/2 are critical values. We proceed with the derivative of *y*.

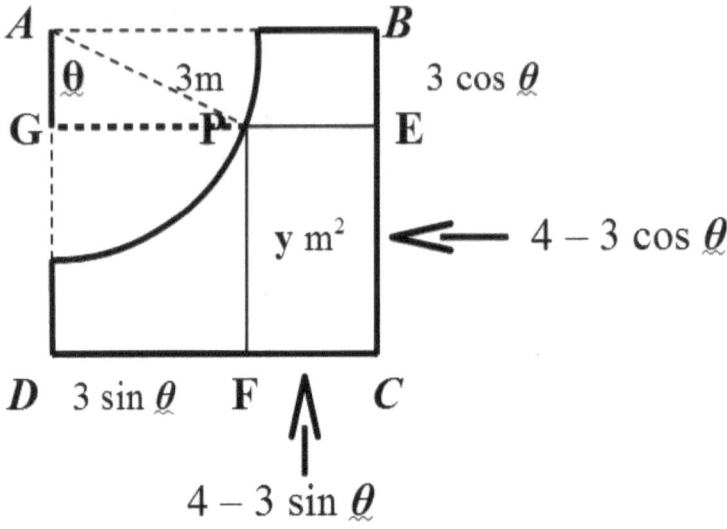

Figure 2

$$\frac{dy}{d\theta} = -3\cos\theta(4-3\cos\theta) + 3\sin\theta(4-3\sin\theta)$$

$$= -12\cos\theta + 9\cos^2\theta + 12\sin\theta - 9\sin^2\theta$$

$$= 12(\sin\theta - \cos\theta) - 9(\sin\theta - \cos\theta)(\sin\theta + \cos\theta)$$

$$= (\sin\theta - \cos\theta)[12 - 9(\sin\theta + \cos\theta)]$$

$$\frac{dy}{d\theta} = 3(\sin\theta - \cos\theta)(4 - 3\sin\theta - 3\cos\theta) \qquad (1)$$

To discover the remaining critical values we solve $dy/d\theta = 0$. This results in the two equations:

$$\tan\theta = 0 \qquad (2)$$

$$4 = 3(\sin\theta + \cos\theta) \qquad (3)$$

Equation (2) leads to $\theta = \pi/4$, and equation (3) involves a sinusoid expression with which we process as follows.

$$\frac{4}{3} = \frac{\sqrt{2}}{\sqrt{2}}(\sin\theta + \cos\theta)$$

$$\frac{4}{3\sqrt{2}} = (\sin\theta)\left(\frac{1}{\sqrt{2}}\right) + (\cos\theta)\left(\frac{1}{\sqrt{2}}\right)$$

$$\frac{4}{3\sqrt{2}} = (\sin\theta)\left(\cos\frac{\pi}{4}\right) + (\cos\theta)\left(\sin\frac{\pi}{4}\right)$$

$$\frac{4}{3\sqrt{2}} = \sin\left(\theta + \frac{\pi}{4}\right)$$

This implies the following:

$$\theta + \frac{\pi}{4} = \sin^{-1}\left(\frac{4}{3\sqrt{2}}\right) \text{ and } \theta + \frac{\pi}{4} = \pi - \sin^{-1}\left(\frac{4}{3\sqrt{2}}\right)$$

$$\theta = \sin^{-1}\left(\frac{4}{3\sqrt{2}}\right) - \frac{\pi}{4} \text{ and } \theta = \frac{3\pi}{4} - \sin^{-1}\left(\frac{4}{3\sqrt{2}}\right)$$

Thus the critical values are as listed below.

$$\left\{0, \ \sin^{-1}\left(\frac{4}{3\sqrt{2}}\right) - \frac{\pi}{4}, \ \frac{\pi}{4}, \ \frac{3\pi}{4} - \sin^{-1}\left(\frac{4}{3\sqrt{2}}\right), \ \frac{\pi}{2}\right\}.$$

(ii) Considering the endpoints 0, and $\pi/2$ we see from equation (1) that

$$y'(0) = 12(-1) + 9(1) = -3 < 0$$
$$y'\left(\frac{\pi}{2}\right) = 12(1) + 9(-1) = 3 > 0$$

From these inequalities we see that the area function's graph falls from its left endpoint and rises to its right endpoint. There are at least a relative if not an absolute maximum value of the area occurring at the endpoints. Evaluating the area at the endpoints,

$$y(0) = y\left(\frac{\pi}{2}\right) = 4\,\mathrm{m}^2$$

We use the second derivative to test the other critical values.

$$y' = -12\cos\theta + 9\cos^2\theta + 12\sin\theta - 9\sin^2\theta$$
$$= 12(\sin\theta - \cos\theta) + 9\cos(2\theta)$$

$$y'' = 12(\cos\theta + \sin\theta) - 18\sin(2\theta)$$

$$y''\left(\tfrac{\pi}{4}\right)=12\left(\sqrt{2}\right)-18(1)<0$$

$$y\left(\tfrac{\pi}{4}\right)=\left(4-\tfrac{3}{\sqrt{2}}\right)\left(4-\tfrac{3}{\sqrt{2}}\right)=\tfrac{41}{2}-12\sqrt{2}\ m^2$$

This value is thus at least a relative maximum.

While it is possible to evaluate the second derivative for

$$\theta=\sin^{-1}\left(\tfrac{4}{3\sqrt{2}}\right)-\tfrac{\pi}{4}$$

and for

$$\theta=\tfrac{3\pi}{4}-\sin^{-1}\left(\tfrac{4}{3\sqrt{2}}\right)$$

by use of a calculator, it is a better challenge to do the evaluation in the old-fashioned manner – exactly. That the value is exactly 2 for these angles can only be appreciated this way. We will only show this for the first angle.

We begin with the Figure 3 diagram of the reference triangle for $\alpha=\sin^{-1}(4/(3\sqrt{2}))$.

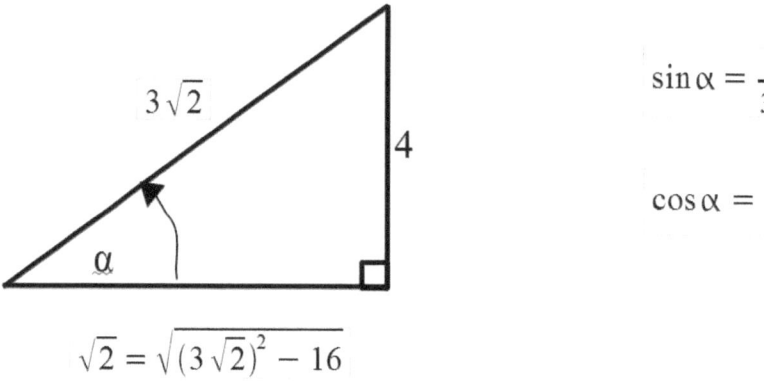

$$\sin\alpha=\tfrac{4}{3\sqrt{2}}$$

$$\cos\alpha=\tfrac{1}{3}$$

$$\sqrt{2}=\sqrt{(3\sqrt{2})^2-16}$$

Figure 3

Using the information from the reference triangle we have the following.

$$y''\left(\alpha-\tfrac{\pi}{4}\right)=12\left(\cos\left(\alpha-\tfrac{\pi}{4}\right)+\sin\left(\alpha-\tfrac{\pi}{4}\right)\right)-36\sin\left(\alpha-\tfrac{\pi}{4}\right)\cos\left(\alpha-\tfrac{\pi}{4}\right) \qquad (4)$$

Now we expand using the formulas for cos(α – π/4) and sin(α - π/4) as follows.

$$\cos\left(\alpha-\frac{\pi}{4}\right)=\frac{1}{3}*\frac{1}{\sqrt{2}}+\frac{4}{3\sqrt{2}}*\frac{1}{\sqrt{2}}$$

$$\sin\left(\alpha-\frac{\pi}{4}\right)=\frac{4}{3\sqrt{2}}*\frac{1}{\sqrt{2}}-\frac{1}{3}*\frac{1}{\sqrt{2}}$$

From these expressions we compute the second derivative from (4) as follows.

$$y''=12\left(\frac{4}{3}\right)-36\left(\frac{2}{3}-\frac{1}{3\sqrt{2}}\right)\left(\frac{2}{3}-\frac{1}{3\sqrt{2}}\right)=16-36\left(\frac{4}{9}-\frac{1}{18}\right)=2>0$$

The second derivative also has the value 2 for the angle $\theta = 3\pi/4 - \sin^{-1}(4/(3\sqrt{2}))$. Thus we conclude that for $\theta = \sin^{-1}(4/(3\sqrt{2}))-\pi/4$ and for $\theta = 3\pi/4 - \sin^{-1}(4/(3\sqrt{2}))$ the area function has a relative minimum.

Looking at the relative maximum values for the area, we compare the value at π/4 with the values for 0 and π/2. The quantity $y(\pi/4) = 41/2 - 12\sqrt{2}$ This is approximately 3.529 which is less than $4 = y(0) = y(\pi/2)$. **Therefore the absolute maximum value of the area is 4 square units for θ = 0 or π/4.**

The absolute minimum value of the area occurs for either of the angles $\theta = \sin^{-1}(4/(3\sqrt{2}))-\pi/4$ or $\theta = 3\pi/4 - \sin^{-1}(4/(3\sqrt{2}))$. We substitute the first one into the area function.

$$y_{min} = y\left(\sin^{-1}\left(\frac{4}{3\sqrt{2}}\right)-\frac{\pi}{4}\right)=$$

$$[4-3\sin\{\sin^{-1}\left(\frac{4}{3\sqrt{2}}\right)-\frac{\pi}{4}\}]\ [4-3\cos\{\sin^{-1}\left(\frac{4}{3\sqrt{2}}\right)-\frac{\pi}{4}\}]$$

$$=[4-3\{\frac{4}{6}-\frac{1}{3}*\frac{1}{\sqrt{2}}\}]\ [4-3\{\frac{1}{3}*\frac{1}{\sqrt{2}}+\frac{4}{6}\}]$$

$$=\left[4-3\left(\frac{2}{3}-\frac{\sqrt{2}}{6}\right)\right]\left[4-3\left(\frac{\sqrt{2}}{6}+\frac{2}{3}\right)\right]$$

$$=\left[2+\frac{1}{\sqrt{2}}\right]\left[2-\frac{1}{\sqrt{2}}\right]$$

$$y_{min} = 3.5 \text{ square units}$$

Let's look at the graph of y (θ).

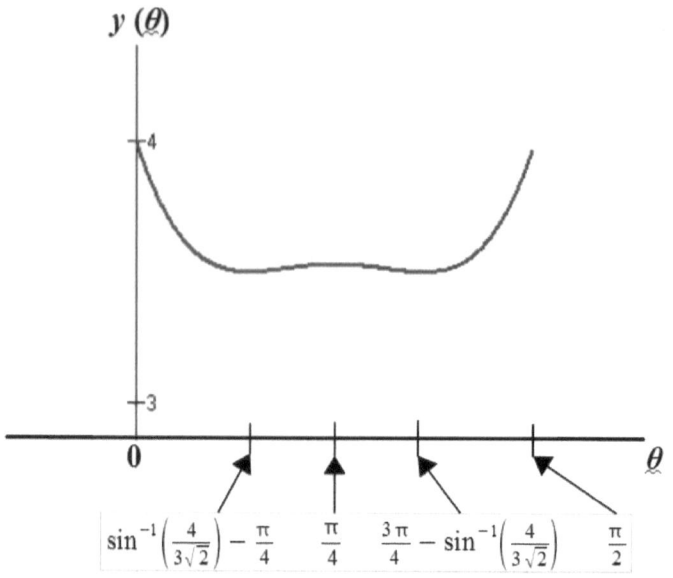

Figure 4

Figure 4 shows the graph of $y(\theta)$. Notice how little variation there is between the extrema as θ varies from $\sin^{-1}(4/(3\sqrt{2})) - \pi/4$ to $3\pi/4 - \sin^{-1}(4/(3\sqrt{2}))$ inclusive.

Note: This article appeared in *Mathematics and Informatics Quarterly* which is an international mathematics journal based in Singapore.

A Related Rates Approach in Determining Galactic Rotation

Students studying science topics such as astronomy are always fascinated with the plethora of information which can be gleaned from a few indirect measurements by applying various mathematical and reasoning processes. This article presents a procedure whereby one can determine the rotation rate of a galaxy by observing the rate of change in two distances – one from the galactic center and the other from a point at the edge of the galaxy's disk. Certain angles are needed to coax out the rotation rate of a galaxy. These can actually be determined from a photo of the subject galaxy. The great Andromeda Galaxy is used as an example.

Introduction

Currently scientists believe that ordinary matter of the type we can see in our telescopes constitutes 20% of the actual matter in the universe. The remaining 80% is termed "dark matter". Included in this 80% is dark energy. One of the more surprising discoveries in modern times is the fact that the expansion rate of the universe is increasing. Before, it was believed that gravity would cause the expansion rate, a relic of the big bang expansion, to gradually slow down.

The presence of dark matter in galaxies has a profound effect on their rotation. Once it was thought that the stars further out from the center rotated at a lesser velocity as distance increased as indicated by the curve labeled A in Figure 1. The actually observed result is that the rotational velocity with distance is remarkably constant as shown in the curve labeled B. Thus to a large extent a galactic disk rotates as a unit.

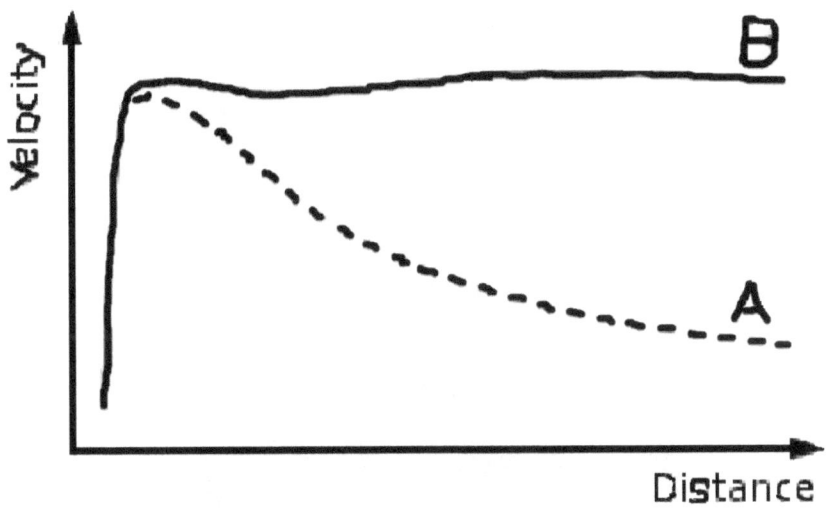

Figure 1
Galaxy rotational velocity is plotted as a function of distance from a galactic center.

An Application of Related Rates

This article presents a method whereby the rotational velocity of a galaxy can be determined using two easily observed velocities combined with angles measurable from a photo. We assume that the visible disk of stars is approximately circular in at least one plane through its center. The spiral galaxy is an important such example of a galaxy which is circular in the plane through its center which is perpendicular to the axis of its rotation.

Figure 2a

Figure 2b

Figure 2c
These galaxies are seen from different angles relative to the plane of their rotations.

Figure 2a shows a photo of a spiral galaxy from a viewpoint in its axis of rotation. Figure 2c shows a side view in which the viewpoint is in the plane of the circular disk of stars. Between these two extremes of perspective the circular disk would appear elliptical like the scene in Figure 2b. By analogous example, the ring system of Saturn would appear circular from a point above a pole, as a line segment from a point above the equator, and as elliptical from a point above 45° north or south latitude.

Figure 3 shows how a circular galactic disk might appear. The boundary indicated by the solid curve appears elliptical but in actuality is not since point B is closer to the observer than point A. In the method to be discussed only two rates need to be known. These are the rate of change in the distance from the observer to the center point, C, and the rate of change in the distance from the observer to point E. These velocities are determined by employing the Doppler Effect whereby a light-

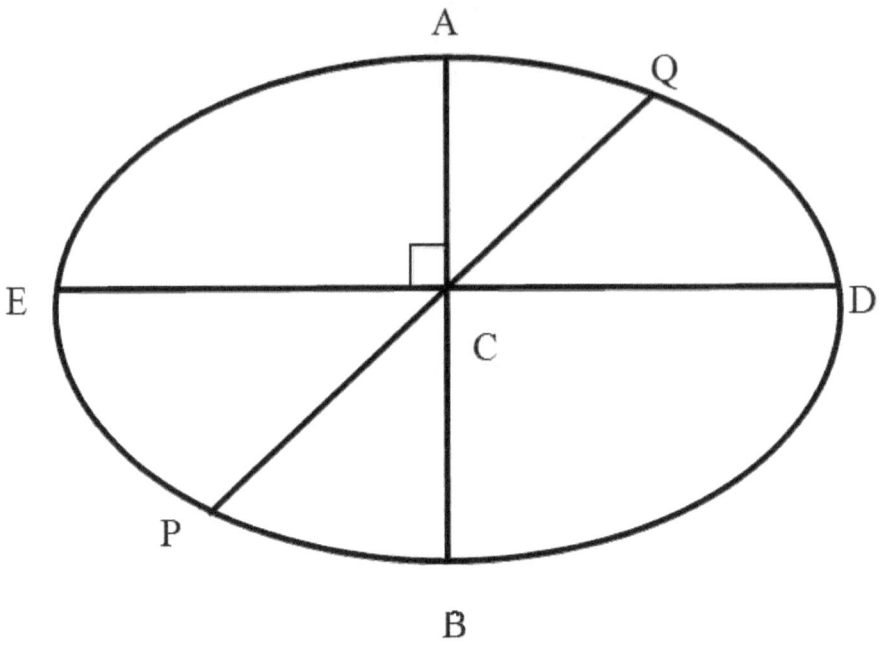

Figure 3.
A circular disk appears as an ellipse when the viewpoint is not in the axis of rotation.

emitting object which is approaching or receding has its light shifted, respectively, toward the violet or red end of the spectrum. It turns out that neither the distance from the observer to point C nor from the observer to point E is necessary to know when accessing the galaxy's rotational velocity. Note that in Figure 3 segment PC looks to be a little longer than segment CQ since from the viewpoint of this figure points P and B are closer than points Q and A. This causes the circle centered at C to appear as an ellipse.

We need to define precisely the locations of points A and E. Look at Figure 3 and segment PQ rotating about its midpoint C. From the observer's perspective point P appears to assume its maximum distance from point C at the position of E and the same maximum at D. Point P assumes two minimum distances from point C – a relative minimum at point B and its absolute minimum distance at point A.

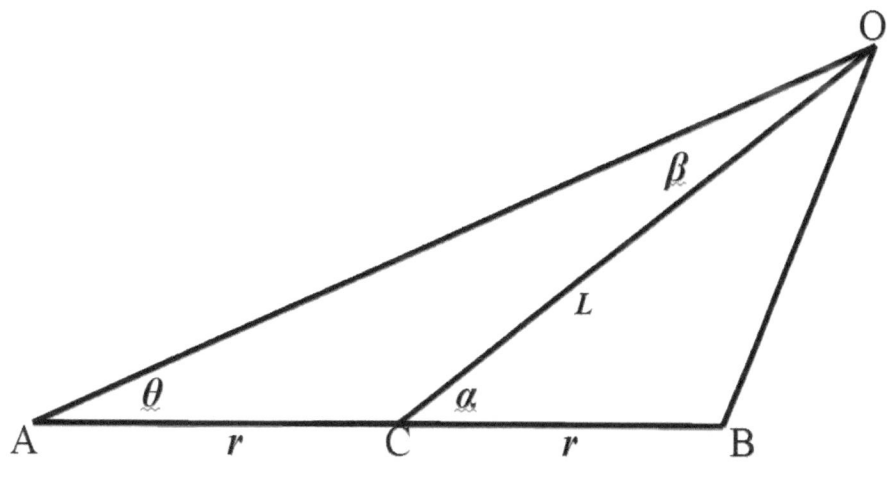

Figure 4

Relative to the viewpoint at O, the three angles shown are important.

In Figure 3 the point from which the observer sees the elliptical view of the circular curve is not shown. Figure 4 is provided both to aid in defining certain quantities and to show the observer's relation to points A and B. Here we see the observer placed at point O. Let L be the length of the line of sight from the observer to point C, the center of rotation. Let α be the acute angle between the line of sight OC and the plane of rotation which contains line segment AB. Let $r = AC = CB$ be the radius of rotation, β be the measure of angle AOC and θ be the measure of angle OAC. Referring back to Figure 1a, 1b, and 1c, we see that in Figure 1a, $\alpha = 90°$, in Figure 1b, α is about 45°, and in Figure 1c, $\alpha = 0°$ since in this case the observer is in the plane of rotation. Note also that since point A appears to be the nearest point to C from the viewpoint of the observer, β is the least measure of the angle between the line of sight to *any* point on the edge of the rotating disk.

In order to determine α, we begin by applying the Law of Sines.

$$\frac{\sin\theta}{L} = \frac{\sin\beta}{r}$$

Since $\alpha = \theta + \beta$, we get $\sin(\alpha - \beta) = \frac{L}{r}(\sin\beta)$, or

$$\alpha = \sin^{-1}\left(\frac{L}{r}(\sin\beta) + \beta\right). \tag{1}$$

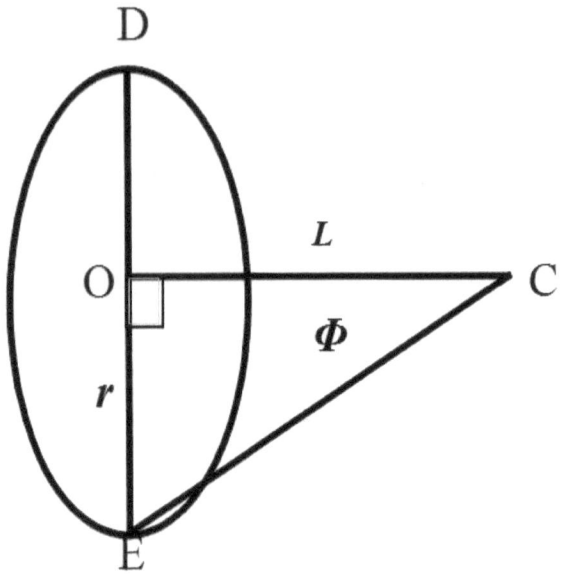

Figure 5

The angle Φ is the maximum angle from the galactic center to a point at the edge of the disk.

Figure 5 shows the location of the observer from a different perspective from that of Figure 4. Let Φ be the *maximum* value of the angle between the line of sight OC and the line of sight from the observer to any point on the edge of the disk. Then by the definition of point E, Φ is the measure of angle COE. Since the line of sight to the center is perpendicular to DE, $\cot \Phi = \dfrac{L}{r}$ and we can express equation (1) as

$$\alpha = \sin^{-1}((\sin \beta)(\cot \Phi)) + \beta, \tag{2}$$

which gives an expression for the acute angle between the line of sight to the center and the plane of points A, B and O, independent of the galactic radius, r.

Our quest is to find an equation relating the rotational velocity of point E about the center at C with two rates of change. One of these is $\dfrac{dL}{dr}$, the rate of change in the distance from the center of the galaxy to the observer, and the other is $\dfrac{ds}{dt}$, the rate of change in the distance between the observer and point E at the edge of the galactic disk. We begin by considering Figure 6 in which an XYZ-coordinate system is established with the center C (0, 0, 0) placed at the origin.

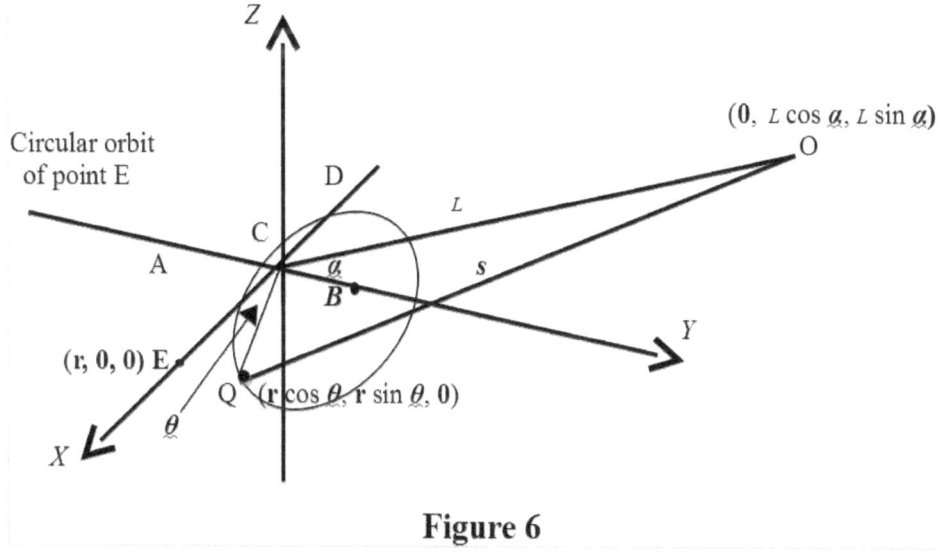

Figure 6

An XYZ-coordinate system is established with the galactic center at the origin.

In Figure 6 point E, at $(r, 0, 0)$, rotates to point Q located at $(r \cos \theta, r \sin \theta, 0)$ where θ is angle ECQ. The observer is at $(0, L \cos \alpha, L \sin \alpha)$ with α is angle OCB. s, a function of time, is the distance from the observer to point E initially, but is shown as OQ since point E rotates to point Q as time elapses. We consider the rotational velocity as being negative if point E is moving toward the observer. At the instant the rate of change in s (that is, ds/dt,) is determined for point E, $\theta = 0$ and point Q is at E. Note that since OC and ED are perpendicular, point O is in the YZ-plane.

From the distance formula we have

$$s^2 = r^2 \cos^2 \theta + (L \cos \alpha - r \sin \theta)^2 + L^2 \sin^2 \alpha$$

which simplifies to

$$s^2 = r^2 - 2rL(\cos \alpha)(\sin \theta) + L^2. \quad (3)$$

Differentiating (3) with respect to time, t, and considering α and r as being constant, yields

$$2s \frac{ds}{dt} = -2r(\cos \alpha)(\sin \theta) \frac{dL}{dt} - 2r L(\cos \alpha)(\cos \theta) \frac{d\theta}{dt} + 2L \frac{dL}{dt}.$$

Solving for $d\theta/dt$ we obtain

113

$$\frac{d\theta}{dt} = \frac{\frac{dL}{dt}[L-r(\cos\alpha)(\sin\theta)] - s\frac{ds}{dt}}{rL(\cos\alpha)(\cos\theta)}$$

which gives the angular velocity of point Q (or E) about C. To obtain the linear velocity, v, of Q in its orbit, we realize that $v = \frac{d\theta}{dt}$, so we have

$$v = \frac{\frac{dL}{dt}[L-r(\cos\alpha)(\sin\theta)] - s\frac{ds}{dt}}{L(\cos\alpha)(\cos\theta)} \tag{4}$$

Since the observation of $\frac{ds}{dt}$ is to be made when Q is at the position of point E, we set $\theta = 0$ in equation (4) to obtain

$$v = \frac{L\frac{dL}{dt} - s\frac{ds}{dt}}{L(\cos\alpha)} \tag{5}$$

which gives the linear velocity of point E about C at the instant $\frac{dL}{dt}$ and $\frac{ds}{dt}$ are observed.

Looking back to Figure 5 we see that $r = L(\tan\Phi)$. Substituting this along with θ into equation (3) we find after simplifying that $s = L(\sec\Phi)$. After replacing s with $L(\sec\Phi)$ in equation (5) and canceling L we have

$$v = (\sec\alpha)\frac{dL}{dt} - \frac{ds}{dt}(\sec\Phi)$$
$$\alpha = \sin^{-1}\{(\sin\beta)(\cot\Phi)\} + \beta \tag{6}$$

The second equation in (6) is a repeat of equation (2).

Figure 7 *The Andromeda Galaxy.*

Using Figure 7, a photo of our nearest major galactic neighbor, the Andromeda Galaxy, we can provide an example illustrating an application of equation (6). The data mentioned below was taken from a picture of the Andromeda Galaxy with the correct aspect ratio. We need to determine the angles Φ and β. These can be found by using a protractor and a ruler.

Hold a protractor perpendicular to the plane of the photo so that the 90° is in contact with the image of the galactic center (C). Thus the straight edge of the protractor is upward since the protractor is upside down. Rotate the protractor so that the straight edge is parallel to the longest axis. With a photo like Figure 7 this axis is from top left to bottom right. Then take a ruler and place one of its ends at the top left point (Point E) and move the ruler until the edge whose end is in contact with the photo passes through the midpoint of the straight edge of the protractor. The angle Φ is determined by obtaining the difference between 90° and the protractor reading where the ruler intersects the curved edge of the protractor. Figure 8 is provided to help illustrate how this works.

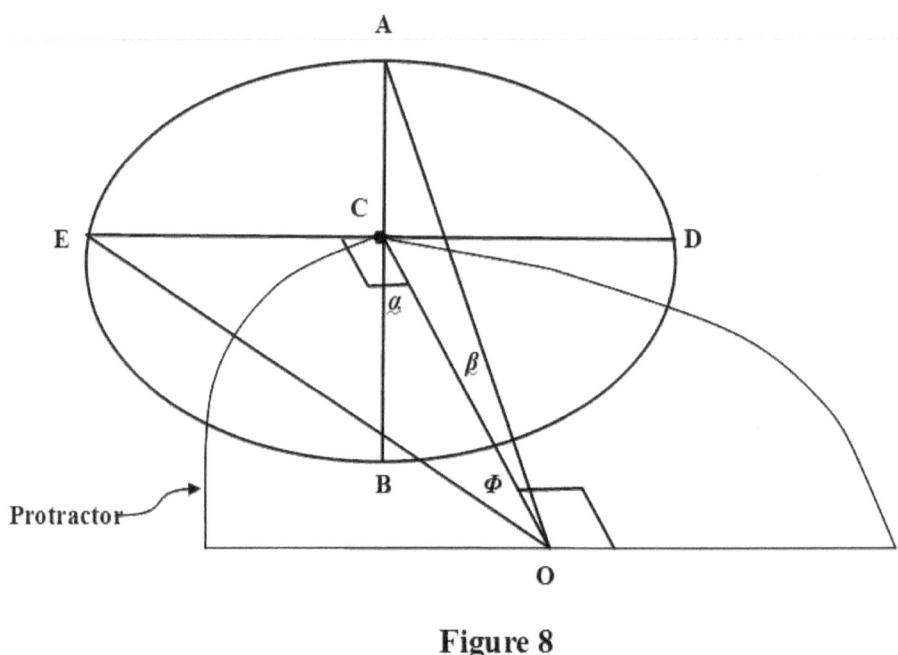

Figure 8

The protractor is shown in the position required to determine Φ.

Figure 8 shows the position of the protractor needed to determine the measure of Φ. Point C is at the 90° position on the protractor and the observer is at the midpoint of the straight edge of the protractor. Segment OC is perpendicular to both segment ED and the protractor's straight edge. The ruler is along segment EO.

To obtain the measure of β repeat the procedure by rotating the protractor until its straight edge is parallel to the shortest axis which is from bottom left to top right in the Figure 7 photo or parallel to AB in Figure 8. Place one end of the ruler at the top right point (A in Figure 8) and let the edge of the ruler intersect the midpoint of the protractor's straight edge (O in Figure 8). The measure of β is the difference between 90° and the reading where the ruler intersects the curved edge of the protractor.

From a printed version of Figure 7 an approximate measure of β is 13.5° and an approximate of measure of Φ is 46.5°. Using equation (2), the value of α is 26.3°. Since equation (6) depends on the specific value of Φ which corresponds to the observed values of $\frac{dL}{dt}$ and $\frac{ds}{dt}$, v would not be correctly determined using 46.5° for Φ. Actual data for the Andromeda Galaxy approximately places β at 0.752° and Φ at 1.743°. This data produces a value of α of 26.3°. The center of the Andromeda Galaxy approaches us at 35 km/s so $\frac{dL}{dt}$ = -35. Also its top left edge, as seen in Figure 7, approaches us at 214.2 km/s giving us $\frac{ds}{dt}$ = -214.2 . These rates are adjusted to account for the sun's rotation within the Milky Way (which, by the way, rotates at 168 miles/sec or 970,000 km/h) and one complete sun circuit requires 225 million years. This is the "galactic year", and by this time measurement, the sun is only 20 galactic years old.

Surprisingly, most people cannot answer the question "What is the largest astronomical object in the sky?" correctly. Most would say that it is our moon since during most solar eclipses the moon's disk completely covers that of the sun. The amazing answer is the Andromeda Galaxy! Look at Figure 9.

Figure 9
The Andromeda Galaxy and the moon are depicted in proper scale in this composite photo.

This is a composite photo showing both the moon and the Andromeda Galaxy. If our eyes could perceive the dim disk of this galaxy it would appear to span several full moons across its longer axis. While Andromeda is a naked eye object, under the best seeing conditions, all one can see is a tiny fuzzy smudge, appearing as a fuzzy star, representing the relatively bright center. Relative to the center of the galaxy, the white circle at 12 o'clock and the white ellipse at 7 o'clock are satellite galaxies of Andromeda. This photo appeared as the December 28, 2006 "Astronomy Picture of the Day and is available at the URL shown below. An interesting picture is published on this site each day. http://antwrp.gsfc.nasa.gov/apod/ap061228.html

Conclusion

A hallmark of the study of astronomy is the use of mathematics in gaining information indirectly. Since we cannot yet escape the confines of the solar system to make direct measurements, mathematics has been a highly utilitarian tool to extend our senses in perceiving remote reality. That mathematics is so successful in describing the properties of the universe has been termed by some people, such as Eugene Wigner, to be a genuine mystery. The following quote from Bertrand Russell sums this up nicely.

> *"Mathematics, rightly viewed, possesses not only truth, but supreme beauty - a beauty cold and austere, like that of sculpture, without appeal to any part of our weaker nature, without the gorgeous trappings of painting or music, yet sublimely pure, and capable of a stern perfection such as only the greatest art can show. The true spirit of delight, the exaltation, the sense of being more than Man, which is the touchstone of the highest excellence, is to be found in mathematics as surely as in poetry."*

We have illustrated an application of mathematics in determining the rotation rate of a tilted disk based on the observed velocities of the center of the disk and a point on its edge combined with angle measurements obtainable from a photograph. The "beauty" and "poetry" of mathematics can be seen in the use of the Law of Cosines and related rates in arriving at a rotational velocity.

Optimizing the Gathering of Solar Energy on the Rooftop

INTRODUCTION

Sometime, perhaps, this country will actually become serious about alternative sources of energy. It is not so much that we are really running out of oil. Rather, we are soon to run out of oil from secure and safe countries where we can depend upon a stable source of supply without having to pay an intolerably high price – both financially and increasingly more important, politically. If our country is not to go the way of previous great civilizations, which reached a zenith, only to decline afterward, we must find widely diverse energy resources, including fossil fuels from our own land. These include wind power, harvesting tidal energy at various seashores, renewable biomass fuels, oil, and solar energy converted to electricity or used directly as a heat source.

While living in Massachusetts, many times the author looked out his bedroom window from his work desk which faced a garage. This was a northern view of that separate building which had an inverted V-shaped slate roof facing both east and west. Depending on the time of day the heat could be seen rising from the east-facing roof in the morning, from both roofs near the noon hour, and from the west-facing roof in the later afternoon hours. This provided the inspiration for the problem situation which is the subject of this article.

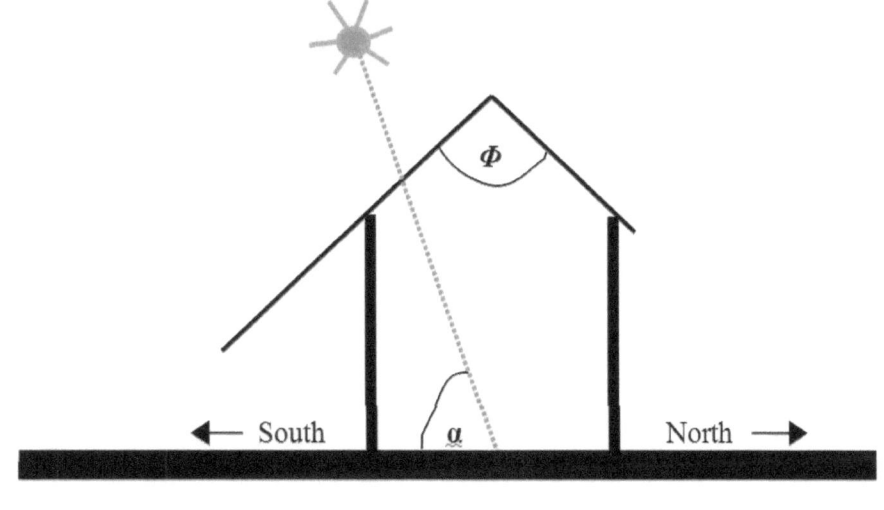

Figure 1

The Problem

A house in the Northern Hemisphere is built with a roof designed to collect solar energy for the heating system. Water is pumped over the roof through a grid of pipes to a storage tank in the basement. For aesthetics, the peak of the roof lies in the plane equidistant from the north and south walls of the house. It is advantageous to use the north roof because during some months the north roof is illuminated, and both roofs are equally efficient during cloudy periods. The south roof, however, has twice the exposed area than the north roof. Let α be the average noon wintertime angle of elevation of the sun, and assume the intensity of heat collected per unit of area is proportional to the square of the sine of the angle of elevation of the sun **with respect to the roof**. (This angle with respect to the roof is not shown in Figure 1.) Let Φ be the angle between the two roofs.

(a) Find Φ as a function of α, where $\alpha \in (0, \frac{\pi}{2})$, and Φ is the angle which *maximizes* the energy received by the entire roof. Prove that the formula devised for Φ does maximize the energy received.

(b) Suppose you own a construction company building these energy efficient homes throughout a geographical region in which α varies from α_1 to α_2 where $\alpha_1 < \alpha_2$ and you want to standardize the roof design so that Φ is the same for all houses built. Assume the building sites are uniformly distributed and that $[\alpha_1, \alpha_2]$ is contained in $(\frac{\pi}{4}, \frac{\pi}{2})$. Write an expression for the best value of Φ so that as a group, all homes maximize receipt of solar energy.

The Solution

(a) Study carefully the angles indicated in the diagram. The angle γ is the angle of elevation of the sun with the north roof and the angle β is the angle of elevation of the sun with the south roof. Let θ be the angle which either roof makes with the ground. While it may seem a little indirect, we will set up a function which gives the intensity of heat received by the entire roof as a function of θ instead of Φ. This approach results in a more straightforward analysis as you shall see. These angles are all indicated in Figure 2.

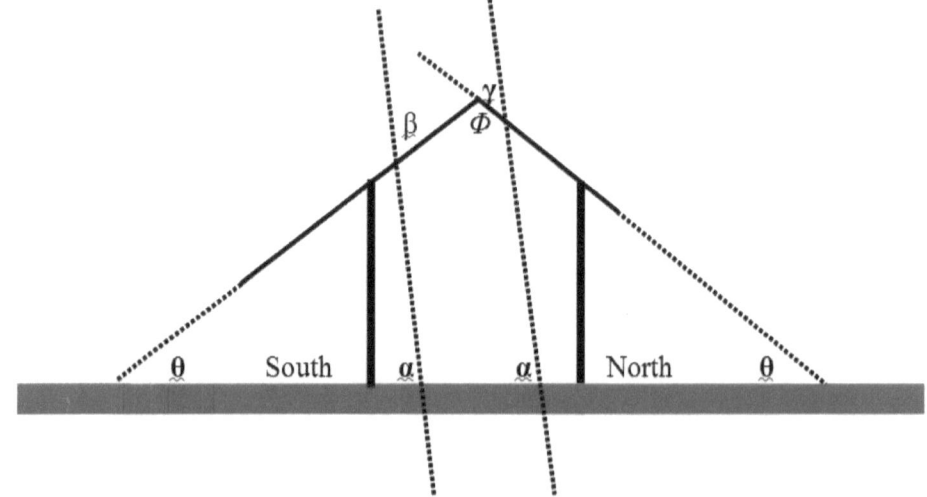

Figure 2

From the big isosceles triangle it is clear that $\Phi = \pi - 2\theta$. Also $\beta = \pi -(\alpha - \theta)$, and $\gamma = \alpha - \theta$. Let **I** be the total intensity of light received by the north and south roofs, and **A** be the area of the north roof. Using **k** as the proportionality constant, we have

$$I = Ak \sin^2\gamma + 2Ak \sin^2\beta$$

$$I = Ak\left(\sin^2(\alpha-\theta)\right) + 2Ak\left(\sin^2\{\pi-[\alpha-\theta]\}\right)$$

$$I = Ak\{2\sin^2(\alpha+\theta) + \sin^2(\alpha-\theta)\}$$

We now differentiate with respect to θ in preparation for setting the derivative equal to zero to find any critical values of θ.

$$\frac{dI}{d\theta} = Ak\{4\sin(\alpha+\theta)\cos(\alpha+\theta) - 2\sin(\alpha-\theta)\cos(\alpha-\theta)\}$$

Now we make use of the double angle formula for the sine function, namely $\sin(2\theta) = 2\sin\theta\cos\theta$.

$$\frac{dI}{d\theta} = Ak\{2\sin(2\alpha+2\theta) - \sin(2\alpha-2\theta)\} = 0$$

$$2\sin(2\alpha+2\theta) = \sin(2\alpha-2\theta)$$

Now we use the sine of the sum and difference formulas on each side and combine like terms, getting

$$3\cos(2\alpha)\sin(2\theta) = \sin(2\alpha)\cos(2\theta)$$

$$\tan(2\theta) = -\frac{1}{3}\tan(2\alpha)$$

$$2\theta = -\tan^{-1}\left(\frac{1}{3}\tan(2\alpha)\right) \tag{1}$$

Since 2θ is positive, the above expression makes sense only if $\frac{1}{3}\tan(2\alpha)$ is a negative number. This means that $2\alpha > \frac{\pi}{2}$ or that $\alpha > \frac{\pi}{4}$. By definition $\Phi = \pi - 2\theta$, so from equation (1) we have

$$\Phi = \pi + \tan^{-1}\left(\frac{1}{3}\tan 2\alpha\right) \text{ if } \alpha > \frac{\pi}{4}$$

as a critical value of Φ. As $\alpha \to \frac{\pi}{4}^+$, $\frac{1}{3}\tan 2\alpha \to -\infty$ which implies that $\tan^{-1}(\frac{1}{3}\tan 2\alpha) \to -\frac{\pi}{2}$ so that $\Phi \to \frac{\pi}{2}$. We also have $\theta \to \frac{\pi}{4}$, $\beta \to \frac{\pi}{2}$, and $\gamma \to 0$. Look at the roof diagram, Figure 2, containing these angles. Geometrically what these tendencies indicate is that as the angle of elevation of the sun, α, nears 45^0, the south roof becomes perpendicular to the sun's rays and the north roof becomes parallel to the sun's rays. For $\alpha < 45^0$, we would keep the south roof perpendicular to the sun's rays while the north roof receives no direct sunlight. Thus β would remain at $\frac{\pi}{2}$ and the relation $\Phi = \Phi - 2\theta = \Phi - 2(\frac{\pi}{2}-\alpha)$ shows that $\Phi = 2\alpha$. Summarizing the complete formula for Φ we have

$$\Phi = \begin{cases} 2\alpha & \text{if } \alpha \in [0, \frac{\pi}{4}] \\ \pi + \tan^{-1}\left(\frac{1}{3}\tan 2\alpha\right), & \text{if } \alpha \in (\frac{\pi}{4}, \frac{\pi}{2}) \end{cases}$$

Justification:

To prove that the above formula for Φ maximizes the amount of heat received by the roof, we need to consider both categories of the formula. For $\Phi = 2\alpha$ there can be no doubt that the intensity, I, is maximized since the larger south roof is perpendicular to the sun's rays. It is the second category which needs a careful analysis, and to justify that we have maximum heat being received, we use the second derivative test.

$$\frac{dI}{d\theta} = Ak[\,2\sin(2\alpha+2\theta) - \sin(2\alpha-2\theta)] = 0$$

$$\frac{d^2I}{d\theta^2} = Ak[\,4\cos(2\alpha+2\theta) + 2\cos(2\alpha-2\theta)]$$

Using the cosine of the sum and difference formulas and simplifying we get

$$\frac{d^2I}{d\theta^2} = 2Ak[\,3(\cos 2\alpha)(\cos 2\theta) - (\sin 2\alpha)(\sin 2\theta)]$$

We must show that this quantity is negative for $2\theta = -\tan^{-1}\left(\frac{1}{3}\tan(2\alpha)\right)$. Since 2α is greater than $\frac{\pi}{2}$, it is clear that $\frac{1}{3}\tan(2\alpha) < 0$ so that

$$-\frac{\pi}{2} < \tan^{-1}\left(\frac{1}{3}\tan(2\alpha)\right) < 0$$

implies that
$$0 < 2\theta < \frac{\pi}{2}.$$
We also know that
$$\frac{\pi}{2} < 2\alpha < \pi.$$

Clearly cos 2θ, sin 2θ, and sin 2α are all positive, and cos 2α is negative so $\dfrac{d^2 I}{d\theta^2} < 0.$

(b) We want the value of Φ which would maximize on average the total energy received by all of the roofs of all homes constructed. It seems appropriate to use the average value of the function, Φ, as the value of its independent variable, α, varies on the interval [α_1, α_2]. The desired value is given by

$$(\Phi)_{avg} = \left(\frac{1}{\alpha_2 - \alpha_1}\right) \int_{\alpha_1}^{\alpha_2} \left\{ \pi + \tan^{-1}\left(\frac{1}{3}\tan 2\alpha\right) \right\} d\alpha$$

A Fluid Flow Application of Linear 2nd Order Differential Equations

Introduction

A Linear Second Order Differential Homogeneous Equation with Constant Coefficients is an equation of the form

$$\frac{d^2 y}{d x^2} + a\left(\frac{d y}{d x}\right) + b y = 0. \tag{1}$$

In the usual differential equations texts one does not find many applications of (1) since such texts are understandably focused on the techniques of solving the various types of equations encountered. However, one commonly found application is the one in which a substance of known concentration is dissolved in a fluid and is pumped into a tank at a given rate. Instantaneous mixing is assumed while the tank is drained. If fluid is removed at the same rate with which it is pumped in, the problem is readily solved by setting up a first order variable separable differential equation. If the rate of outflow is not the same as the rate of outflow, the resulting differential equation is of the form

$$\frac{d S}{d t} + P(t)S = K$$

which can be solved by use of the integrating factor $\rho = e^{\int p(t)dt}$. The subject of this article is an application devised by the author in which a system of fluid flow involving three tanks eventually results in an equation of the form of (1).

The Problem

Consider the system of tanks shown in Figure 1. (After the end of this paragraph you will want to use the 5-way controller to magnify Figure 1.) Tank A initially contains 100 gallons of brine in which 100 pounds of salt are dissolved. Tank B initially contains 100 gallons of pure water. Pure water flows into Tank A at the constant rate of 3 gallons per minute and the mixture (assuming instantaneous mixing) flows into Tank B at the rate of 4 gallons per minute. Meanwhile from Tank B the mixture is pumped back into Tank A at the rate of 1 gallon per minute. At the same time 3 gallons per minute of mixture from Tank B is pumped into Tank C. Determine the maximum amount of salt in Tank B.

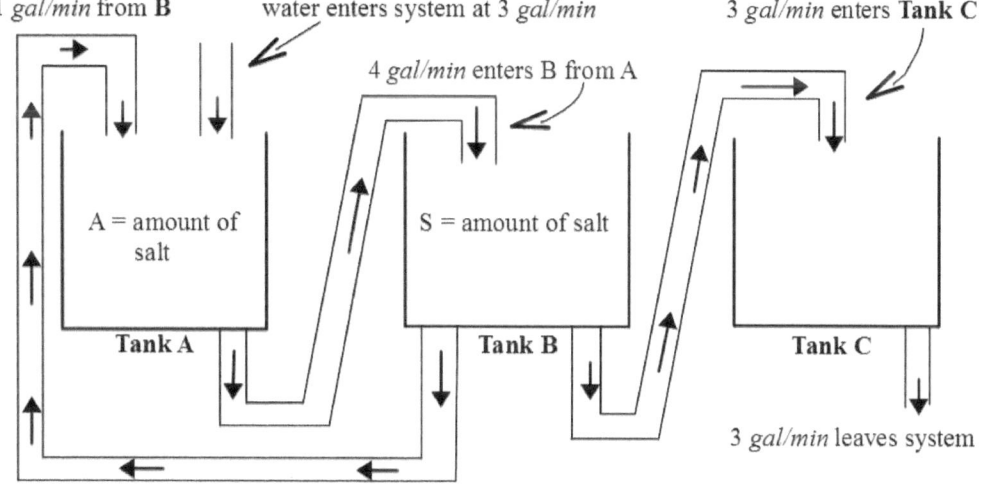

Figure 1

Solution

Let A = the amount of salt in Tank A and S = the amount of salt in Tank B. $\frac{A}{100}$ = the number of pounds per gallon present in Tank A at any time, t. Since the outflow rate from Tank A is – 4 gallons per minute, $\left(\frac{A}{100}\right)(-4) = -\frac{A}{25}$ pounds per minute is the outflow rate for the salt in Tank A. Similarly $\left(\frac{S}{100}\right)(+4) = \frac{S}{100}$ pounds per minute is the inflow rate of salt into tank A. Thus we have

$$\frac{dA}{dt} = -\frac{A}{25} + \frac{S}{100} \tag{2}$$

as a complete description of the instantaneous rate of change of the salt content in Tank A. Since the outflow from Tank A goes directly into Tank B, A/25 pounds per minute is the rate of inflow into Tank B. Since a total of 4 gallons per minute flow out of Tank B, the rate of outflow of salt from Tank B is given by $\left(\frac{S}{100}\right)(-4) = -\frac{S}{25}$ pounds per minute. This results in

$$\frac{dS}{dt} = \frac{A}{25} + -\frac{S}{25}. \tag{3}$$

Beginning the process of solving the problem we add (2) to (3) to eliminate A leading to equation (4).

125

$$\frac{dA}{dt} = -\frac{3S}{100} - \frac{dS}{dt} \qquad (4)$$

Differentiating both sides of (3) yields

$$\frac{d^2S}{dt^2} = \frac{1}{25}\left(\frac{dA}{dt}\right) - \frac{1}{25}\left(\frac{dS}{dt}\right) \qquad (5)$$

We eliminate A by substituting the expression for $\frac{dA}{dt}$ provided by equation (4) into equation (5). This leads to

$$\frac{d^2S}{dt^2} + \frac{2}{25}\left(\frac{dS}{dt}\right) + \frac{3}{2500}(S) = 0 \qquad (6)$$

which is a linear second order differential homogeneous equation with constant coefficients.

The solution of (6) is facilitated by solving the characteristic equation $r^2 + \frac{2}{25}r + \frac{3}{2500} = 0$.

From

$$50^2 r^2 + 200 r + 3 = 0 = (50r + 3)(50r + 1)$$

we obtain $r_1 = -3/50$, $r_2 = -1/50$. The general solution is $S(t) = C_1 e^{-3t/50} + C_2 e^{-t/50}$. Since $S(0) = 0$, $0 = C_1 + C_2$. We put $-C_1 = C_2$ so that the solution becomes

$$S(t) = C_1\left(e^{-3t/50} - e^{-t/50}\right). \qquad (7)$$

Since we are seeking a maximum of S we differentiate (7) and set the result equal to zero.

$$\frac{dS}{dt} = C_1\left(\frac{-3}{50} e^{-3t/50} + \frac{1}{50} e^{-t/50}\right) \qquad (8)$$

Since $\frac{dS}{dt} = 0$, $3 = e^{t/25}$ implies that $t = 25(\ln(3))$.

From (7) we get the maximum value

$$S(25 \ln 3) = C_1\left(e^{-(3\ln 3)/2} - e^{-(\ln 3)/2}\right)$$

which, after simplifying, can be written

$$S_{max} = C_1\left(\frac{-2}{3\sqrt{3}}\right).$$

To complete the solution we need the value of C_1. Note that we have not yet used the fact that at $t=0$, $A=100$. Look back to equation (3). At $t=0$, $\frac{dS}{dt} = \frac{100}{25} - \frac{0}{25} = 4$. Thus using equation (8) we have

$$4 = C_1\left(\frac{-3}{50}e^0 + \frac{1}{50}e^0\right) = C_1\left(\frac{-1}{25}\right)$$

so that $C_1 = -100$. Finally we conclude $S_{max} = -100\left(\frac{-2}{3\sqrt{3}}\right) = 38.5$ pounds, approximately.

Note that from (8) the second derivative is given by

$$\frac{d^2S}{dt^2} = C_1\left(\frac{9}{50^2}e^{-3t/50} - \frac{1}{50^2}e^{-t/50}\right).$$

After simplifying we see that

$$S''(25\ln 3) = \frac{C_1}{50^2}\left(\frac{3\sqrt{3}-1}{3}\right)$$

which is clearly negative for $C_1 = -100$. Thus the maximum is verified.

Time Dilation Explained

A major breakthrough by Albert Einstein was the realization that light travels at a constant speed. This had major consequences for physics. One which has captured people's imagination is time dilation which is the fact that time is relative to each observer. If a very rapidly moving observer's watch measures a time interval of ten minutes, for example, many hours of time could elapse for a stationary observer (stationary relative to the moving observer). The famous Twin Paradox describes a pair of twins, one of which leaves earth as a youth and moves at nearly the speed of light for his voyage. Upon his return to earth, his twin brother is an old man while the traveler has hardly aged at all!

The math behind this phenomenon is relatively (no pun intended) easy to explain as the underlying principles are the $D = R*T$ formula and the Pythagorean Theorem.

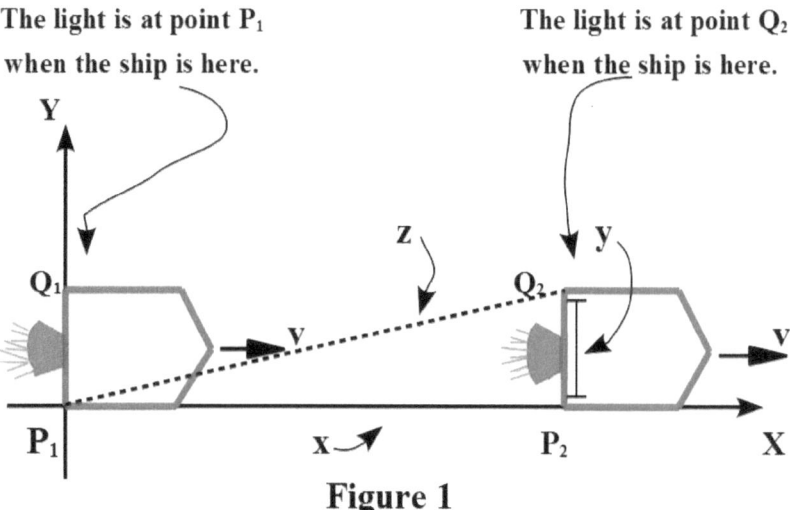

Figure 1

Consider the Figure 1 *xy*-coordinate system as being stationary with only the stylishly drawn spaceship moving to the right. Suppose a beam of light is sent across the spaceship from point P to point Q. We show below that, since the speed of light must be perceived as being constant for *any* observer, that time slows down for a moving observer as measured by the stationary observer. For example, if the passenger's watch measures a 10 second time interval, the stationary person's watch could measure one minute!

Let v = the speed of the spaceship, c = the speed of light. We have a light beam which moves from point P to point Q and two observers, one in the spaceship and one stationary who is watching the action. The observer in the spaceship times the movement of light from point P to point Q and finds that the time required is T_0 for its travel through the distance, y. This means that $y = (c)(T_0)$.

The stationary observer watches the spaceship move from P_1P_2, a distance of x, in the time T. Thus we have $x = (v)(T)$. The stationary observer also perceives the light to move from point P_1 to point Q_2, which is a distance of z, so that $z = (c)(T)$. To compare the time T with T_0 we use the Pythagorean Theorem on triangle $P_1Q_2P_2$ to obtain

$$z^2 = x^2 + y^2$$

$$c^2 T^2 = v^2 T^2 + c^2 T_0^2$$

Solving for T yields

$$T = \frac{T_0}{\sqrt{1 - \frac{v^2}{c^2}}} \tag{1}$$

Since $v < c$ it is clear that $T > T_0$ so that time passes more slowly for the observer in the spaceship **relative to the stationary observer.**

Equation (1) allows one to describe time dilation for specific examples. For example, suppose an astronaut embarks on a space mission, traveling on a round trip to Alpha Centauri which is approximately 4.4 light years away. Alpha Centauri is the nearest star system, consisting of three stars, Alpha Centauri A, B, and Proxima Centauri. The A star is similar to our sun and would be an interesting object for close-up study if and when we master traveling speeds which are a sizable fraction of the speed of light.

Let's say one of a pair of twins sets off to Alpha Centauri averaging half of the speed of light (a rate of 0.5c). It would require 8.8 years to get to the system and 8.8 years to return for a total of 17.6 years. This would be the elapsed time for the trip experienced by the earth-bound twin waiting for his brother's return. Thus in equation (1) we have $T = 17.6$ with $v = 0.5c$. Substituting into (1) yields

$$17.6 = \frac{T_0}{\sqrt{1-(.5c/c)^2}} = \frac{T_0}{\sqrt{3/4}}$$

which leads to

$$T_0 = 17.6 \left(\frac{\sqrt{3}}{2}\right) \approx 15.24 .$$

Therefore while the earth-bound twin waits about 17 years and 7 months, upon his return, the astronaut brother perceives the trip to have taken 15 years and 3 months. The time dilation causes the 2 years and 4 months difference.

The Weight Watcher Function

Introduction

What is the weight watcher function? Starvation! This is a fine one word answer to the question. But you are not satisfied with this at all. You want a mathematical analysis of this, no doubt. You have navigated to just the right location. We will answer this question with a formula which, based on a person's caloric intake and activity level, will predict a person's weight as a function of the time while on a certain diet.

You first need some background. We all require calories in our food. Calories are a measure of the energy supplied and are a measure of heat. Just set fire to that chocolate cookie, assuming that its water content has been removed. Then measure the heat released and you will know the caloric value. Of course, a fire is a "fast oxidation" process. When we eat, the body releases the energy in food with a process called "slow oxidation". The two types of oxidation are similar in that heat is released.

It is very important for our purposes here to realize that for every 3500 calorie surplus, the body stores this excess as one pound of body fat. The amount of calories one needs to maintain constant weight varies with activity level. Generally one requires from 16 to 19 calories per pound of body weight to maintain constant weight. We refer to this as the daily allotment of calories per pound.

Definition of the Variables and Constants

The following variables and constants will be employed in this discussion.

a = the daily allotment of calories per pound of body weight for constant weight maintenance. This is a constant between 16 and 19.

N = the daily number of calories consumed on a particular diet. This is a constant.

w_0 = a person's initial weight. This is a constant measured at the beginning of a diet.

t = the number of days a diet is followed. This is the independent variable.

w_t = the weight after t days. This is the dependent variable.

A Basic Assumption

Assuming a person weighs $w0$ pounds, then that person needs aw_0 calories at that time to maintain a constant weight as the energy intake balances the energy outflow due to body temperature maintenance and exercise level. Since we will be assuming a person on a diet is losing weight, we must assume that $(aw_0 - N) > 0$ since the number of calories consumed, N, is less than the number needed for constant weight.

Development of the Weight Watcher Function

Let the sequence $w_1, w_2, w_3, \cdots$ represent the sequence of weights on day 1, 2, 3, ... of the diet. Since we have assumed that $(aw_0 - N) > 0$, $(N - aw_0)$ is negative. Knowing that it takes a 3500 calorie deficit to result in a one pound loss, the negative number, $\dfrac{N - aw_0}{3500}$, is the actual weight loss to occur on day #1 of the diet. Thus

$$w_1 = w_0 + \frac{N - aw_0}{3500}$$

$$w_1 = \left(1 - \frac{a}{3500}\right)w_0 + \left(\frac{N}{3500}\right)$$

For convenience we define the constants

$$A = \left(1 - \frac{a}{3500}\right), \quad B = \frac{N}{3500}.$$

Thus $w_1 = A w_0 + B$. Since the weight on the second day is derived in **exactly the same way** as the weight was obtained on the first day, we can write

$$w_2 = A w_1 + B$$

Continuing in the same was, we can form a general statement producing the weight on day t from the weight on day t-1 as follows:

$$w_t = A w_{t-1} + B \tag{1}$$

We now turn our attention to using this procedure to derive a formula for w_t in terms of the constants A, B, w_0, and the variable, t. Study the sequence of relationships below.

$$\begin{aligned}
w_1 &= A w_0 + B \\
w_2 &= A w_1 + B = A(A w_0 + B) + B \\
&= A^2 w_0 + B(A + 1) \\
w_3 &= A w_2 + B = A\{A^2 w_0 + B(A + 1)\} + B \\
w_3 &= A^3 w_0 + B(A^2 + A + 1) \\
w_4 &= A^4 w_0 + B(A^3 + A^2 + A + 1) \\
&\cdots \\
w_t &= A^t w_0 + B\left(A^{t-1} + A^{t-2} + \cdots + A^3 + A^2 + A + 1\right)
\end{aligned}$$

More compactly, we can express w_t with sigma notation.

$$w_t = A^t w_0 + B \sum_{j=0}^{t-1} A^j \tag{2}$$

Notice that

$$\sum_{j=0}^{t-1} A^j = 1 + A + A^2 + A^3 + \cdots A^{t-1}$$

is a geometric series containing t number of terms in which A is the constant ratio. Using the geometric sum formula $S_n = \dfrac{a_1(1-r^n)}{1-r}$ we get

$$\sum_{j=0}^{t-1} A^j = \frac{1(1-A^t)}{1-A} = \frac{A^t - 1}{A - 1}$$

Substituting this into (2) the result is

$$w_t = A^t w_0 + B \left(\frac{A^t - 1}{A - 1} \right)$$

$$= A^t w_0 + A^t \left(\frac{B}{A-1} \right) - \frac{B}{A-1}$$

$$W_t = A^t \left(w_0 + \frac{B}{A-1} \right) - \frac{B}{A-1}$$

where $A = \left(1 - \dfrac{a}{3500}\right)$, and $B = \dfrac{N}{3500}$. Substituting for A and B we arrive at our goal, the Weight Watcher Function. (Viewing the next equation requires landscape mode.)

$$w_t = \left(1 - \frac{a}{3500}\right)^t \left[w_0 + \frac{\frac{N}{3500}}{\frac{-a}{3500}} \right] - \left(\frac{\frac{N}{3500}}{\frac{-a}{3500}} \right)$$

$$= \left(\frac{3500 - a}{3500} \right)^t \left(w_0 - \frac{N}{a} \right) + \frac{N}{a}$$

$$w_t = \frac{1}{a}\left\{ N + (a w_0 - N)\left(\frac{3500-a}{3500} \right)^t \right\} \tag{3}$$

An Example

A 200 pound man decides to go on a 2500 calorie per day diet. (Assume a = 16 calories/pound)
(a) Determine this person's weight watcher function.
(b) Using the customized function for this person found in (a), predict when his weight will reach 175 pounds.

133

Solution:
(a) a = 16, w_0 = 200, N = 2500. aw_0 = 3200 < 3500 so weight loss will occur.

$$w_t = \frac{1}{16}\left\{2500 + 700\left(\frac{174.2}{175}\right)^t\right\}$$

(b) We need to solve the equation

$$175 = \frac{1}{16}\left\{2500 + 700\left(\frac{174.2}{175}\right)^t\right\}$$

for t. Simplifying the numbers we have $\frac{3}{7} = \left(\frac{174.2}{175}\right)^t$. Taking the logarithm of both sides (base 10 or base e), we find

$$\frac{\log 3 - \log 7}{\log 174.2 - \log 175} = t \approx 185 \text{ days}$$

The Rest of the Story!

Normally what you just read would be the whole story. When originally developed the formula had to be tested. This was done with real people – teachers at Wellesley High School. The results were favorable, but with nowhere to look up the formula devised above, a way to get confirmation of the formula was available by deriving it by means of a completely different mathematical approach. **THAT** is the rest of the story which involves calculus!

Clearly, weight loss should be more rapid the greater the calorie deficit. It should be twice as rapid if the deficit in calories could itself be doubled. Does this seem reasonable? We are really saying here is that **the rate of weight loss is proportional to the deficit in calories.** Analytically, if we let W = f (t) be the weight as a function of time, the statement in italics becomes

$$\frac{dW}{dt} = k(aW - N) \tag{4}$$

The important question here is this: Is differential equation (4) equivalent to equation (3)? This question is much more focused as: **Is equation (3) a solution of (4)?** We will find out!

In order to solve (4) properly we need two data points – two sets of corresponding t and W values. Let W(0) = w_0 and W(1) = $w_0 + \frac{N - aw_0}{3500}$. We begin as usual with the separation of variables in (4).

$$\int \frac{dW}{aW - N} = \int k\, dt$$

$$kt = \frac{1}{a}\ln(aW - N) + C$$

134

If $t = 0$ and $W = w_0$ we can evaluate C. $C = -\frac{1}{a}\ln(aw_0 - N)$ Substituting and combining the log terms we get

$$kt = \frac{1}{a}\ln\left(\frac{aW - N}{aw_0 - N}\right) \tag{5}$$

To evaluate k we use the second data point: when $t = 1$, $W = w_0 + \frac{N - aw_0}{3500}$.

$$k = \frac{1}{a}\ln\left(\frac{a\left(w_0 + \frac{N - aw_0}{3500}\right) - N}{aw_0 - N}\right)$$

Obviously this cries out for a little simplification! We begin by eliminating the complex fraction.

$$k = \frac{1}{a}\ln\left(\frac{3500\,aw_0 + aN - a^2 w_0 - 3500N}{3500(aw_0 - N)}\right)$$

$$k = \frac{1}{a}\ln\left(\frac{(aw_0 - N)(3500 - a)}{3500(aw_0 - N)}\right)$$

$$k = \frac{1}{a}\ln\left(\frac{3500 - a}{3500}\right)$$

Substituting for k in equation (5) results in the following.

$$t\left(\frac{1}{a}\ln\left(\frac{3500 - a}{3500}\right)\right) = \frac{1}{a}\ln\left(\frac{aW - N}{aw_0 - N}\right)$$

$$\ln\left(\frac{3500 - a}{3500}\right)^t = \ln\left(\frac{aW - N}{aw_0 - N}\right)$$

$$\left(\frac{3500 - a}{3500}\right)^t = \left(\frac{aW - N}{aw_0 - N}\right)$$

$$(aw_0 - N)\left(\frac{3500 - a}{3500}\right)^t = aW - N$$

$$W(t) = \frac{1}{a}\left\{N + (aw_0 - N)\left(\frac{3500 - a}{3500}\right)^t\right\} \tag{6}$$

Indeed, equation (6), derived from differential equation (4) precisely matches equation (3), derived from equation (1).

Look at equation (6) again. If t approaches infinity, $W(t)$ approaches N/a. This shows that if consumption is unchanged over time, the weight tends toward a definite minimum limit, and the rate of weight loss tapers off. This is a primary factor causing some people to give up on a diet. They observe the most rapid loss rate at the beginning with a discouraging slow down in the loss rate later. The graph in Figure 1 shows this phenomenon.

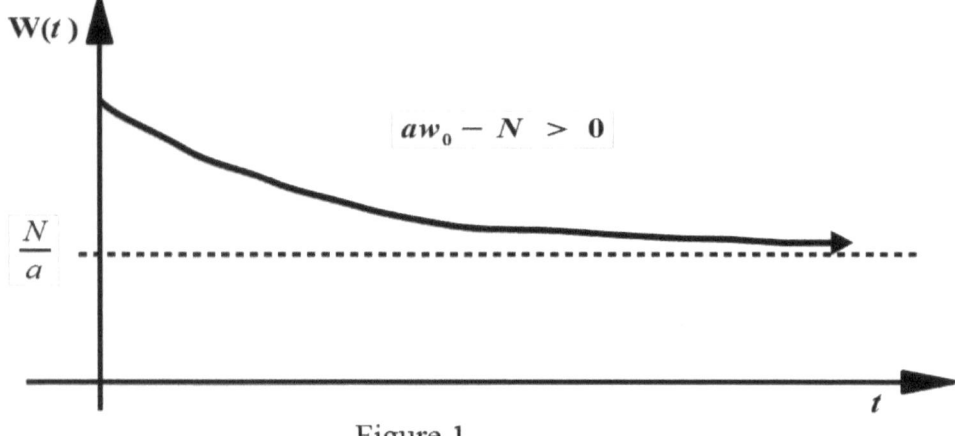

Figure 1

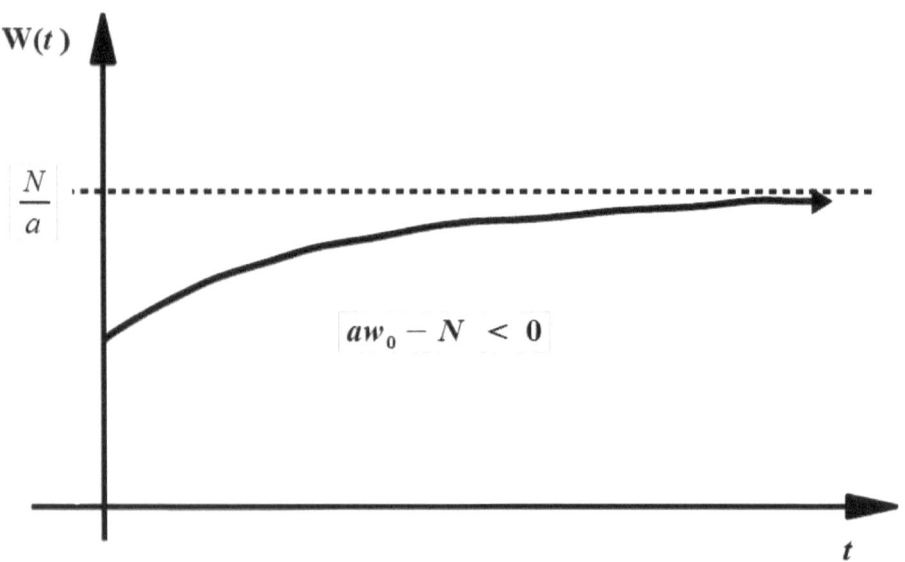

Figure 2

In the case that $aw_0 - N < 0$ there would be weight gain! In this event a person's weight would rise most rapidly at first, then level off at the $\frac{N}{a}$ level. Figure 2 shows the weight gain situation over time.

www.ingramcontent.com/pod-product-compliance
Lightning Source LLC
Chambersburg PA
CBHW021825170526
45157CB00007B/2689